Fluid Machinery

IMechE
Conference Transactions

I MECH E

Seventh European Congress on

Fluid Machinery
For the Oil, Petrochemical, and Related Industries

15–16 April 1999
The Netherlands Congress Centre, The Hague, The Netherlands

Organized by the Fluid Machinery Committee of the Power Industries Division of the Institution of Mechanical Engineers (IMechE)

Sponsored by

GHH Borsig
John Crane
Sulzer Turbo
Fluent Europe Limited
AMEC Process and Energy
Sundstrand Compressors

Co-sponsored by

EEMUA
KIVI
VDI

IMechE Conference Transactions 1999–2

**Professional
Engineering
Publishing**

Published by Professional Engineering Publishing Limited for The Institution of Mechanical Engineers, Bury St Edmunds and London, UK.

First Published 1999

ISSN 1356–1448
ISBN 1 86058 217 6

A CIP catalogue record for this book is available from the British Library.

Printed by The Book Company, Suffolk, UK.

Congress Organizing Committee

R Bouwman
Berkhof Jonckheere

R Elder
Cranfield University

J Glasgow (Chairman)
AMEC Process and Energy Limited

G Habets
Shell International Oil Products

G D Johnson
Dresser-Rand Company

R Palgrave
David Brown-Union Pumps Limited

D Redpath
BP Oil

I Rhodes
BHR Group Limited

G Robson
Ingersoll-Dresser Pumps (UK) Limited

Related Titles of Interest

Title	Editor/Author	ISBN
Subsea Control and Data Acquisition	L Adriaansen, R Phillips, C Rees, and J Cattanach	0 85298 993 8
Fluid Machinery for the Oil, Petrochemical, and Related Industries	IMechE Conference 1996–4	0 85298 994 6
Computational Fluid Dynamics in Fluid Machinery	IMechE Seminar 1998–13	1 86058 165 X
Pumping Sludge and Slurry	IMechE Seminar 1998–1	1 86058 155 2
Multiphase Technology from the Arctic to the Tropics (BHRG No 31)	J P Brill and G A Gregory	1 86058 139 0
Protecting the Environment In Process Industry (BHRG No 30)	J A R Muir	1 86058 138 2

For the full range of titles published by Professional Engineering Publishing contact:

Sales Department
Professional Engineering Publishing Limited
Northgate Avenue
Bury St Edmunds
Suffolk
IP32 6BW
UK

Tel: 01284 724384
Fax: 01284 718692

Contents

Session 6

Session 7

Session 8

Additional Papers

Authors' Index

C556/008/99

Development of a hydraulic turbine in liquefied natural gas

G L G M HABETS
Shell International Oil Products BV, The Hague, The Netherlands
H KIMMEL
Ebara International Corporation, Cryodynamics Division, Nevada, USA

1.0   SYNOPSIS

This paper addresses the development of a submerged hydraulic turbine generator in the LNG Complex for the Sultanate of Oman. The concept is based on the utilisation of Liquefied Natural Gas under pressure that expands over a hydraulic turbine consisting of several centrifugal stages. The expanding liquid is cooled in the expansion process during which the temperature decreases while the pressure is relieved. The process has attractive thermodynamics effectively utilised to drive a directly coupled generator. The entire unit is submerged in the cryogenic fluid and is contained in a pressurised vessel.

The paper covers the development of such a novel hydraulic turbine, the mechanical concept of which is based on submerged liquid pumps in cryogenic service. It demonstrates that this was carried out in close co-operation between the end user and the turbine manufacturer. Particularly many novel aspects required in-depth analysis in order to minimise the risk and to comply with the strict requirements for new equipment development imposed by the end user. The best possible thermodynamic efficiency had to be achieved satisfying not one but a range of operating conditions. This efficiency had to be confirmed both through theoretical and empirical means.

Due to the nature of variable speed of the unit the mechanical analysis to satisfy rotordynamic aspects and the electrical requirements to comply with grid frequency limitations required close attention. Described are various steps and verification points that were agreed at the beginning of the development process. By means of thorough testing of all units at the manufacturers works utilising cryogenic LNG as imposed by the end user, infant mortality problems could be eliminated and full acceptance realised prior to shipment of the units to their destination in Sur, Oman.

It is illustrated that development of new equipment can be achieved through close communication between end user and manufacturer resulting in a novel piece of equipment while the risks have minimised by means of stringent verification steps.

2.0   INTRODUCTION

2.1 Role of a hydraulic turbine in large LNG plants
The Oman LNG complex will consist of two LNG trains with capacity of 3.1 million tons per annum. The process is based on seawater cooling and Propane and Mixed Refrigerant cooling cycle. The Propane cycle contains a General Electric Frame 6B type driven compressor and the MR cycle compressors are driven by a General Electric Frame 7EA gasturbine.

The treated natural gas is condensed and subcooled in a Main Cryogenic Heat Exchanger. The final Liquefied Natural Gas under pressure is expanded before it is forwarded to large atmospheric pressure LNG storage tanks. This pressure letdown is achieved through hydraulic expanders in parallel with Joules-Thompson expansion valves that act as fall back. Heavy MR liquid is also expanded via a separate hydraulic turbine. Figure 1 indicates the hydraulic turbines within the overall plant process scheme.

2.2 History
The concept of replacing the rather inefficient pressure letdown through JT valves by equipment that allows a more efficient expansion of the liquid with improved efficiency is conventional. The first application of hydraulic turbines in a LNG complex dates from 1995.These units utilise variable inlet geometry nozzles, wicket gates, and deliver electric power at fixed speed through an external generator to the plant electrical grid. Such units have been in operation in a Malaysian LNG plant since 1995. Initial discussions on the concept of a submerged hydraulic turbine took place in April 1996. A prototype of a single stage unit was demonstrated late 1996. (1)   With the completion of the manufacturers tests of the novel type Oman LNG units in August 1998, two manufacturers with a capacity for providing hydraulic turbines for the LNG process became available. These are Flowserve (Byron Jackson) with a variable geometry hydraulic turbine and EBARA International Corporation with a variable speed hydraulic turbine.

2.3 Operational requirements
The main parameters in the design of LNG plants are the composition of the feed gas and the type of cooling selected. Apart from the type of process that is selected for cooling also the prime rejection of heat to the environment, air-cooling or water-cooling is important. This results in various operating cases that are the basic requirements for equipment selection and design. In order to obtain maximum plant capacity at lowest costs the efficiency of equipment and in particular hydraulic turbines is essential. Not only a high efficiency at design capacity but also an efficiency characteristic that remains at high values and constant over the expected flow ranges is important. (2)

C556/008 © IMechE 1999

The next significant operational requirement is a high reliability. Equipment shall be of robust design with proven experience. Although the hydraulic turbine is backed up with a J-T valve, production is reduced due to lower efficiency of the J-T valve.

Safety requirements demand a high technical integrity of equipment that contains hydrocarbons and in particular pressure containing components thereof. Fully submerged units, mounted inside a pressure vessel without the need for shaft seals appear attractive from this point of view.

3.0   TURBINE DESCRIPTION

3.1 General features
The concept is based on the submerged cryogenic pumps that have been applied numerous times in cryogenic processes. The turbine consists of several stages. The flow enters through a nozzle ring with fixed geometry radially into the turbine runners. Refer figure 2. The generator is submerged in the LNG and directly coupled to the turbine shaft. The fluid enters along the generator casing while a small portion is diverted through the generator rotor/stator gap to effect cooling. The shaft is supported by rolling element bearings at the top and bottom of the generator and at the bottom of the turbine. Axial thrust is balanced by means of a hydraulic balancing self-adjusting variable gap mechanism.

A main feature is the variable speed. The power generated in the generator is supplied to a Variable Speed Constant Frequency System (VSCF), which converts the frequency of the power supply corresponding with the turbine running speed to the desired grid frequency. Refer figure 3.

3.2 Performance
A typical performance characteristic is shown in figure 4. The operating window typically has two boundary lines: the zero torque line and the zero speed line. The zero torque line is the free running mode of the turbine, where no power is extracted from the generator. All hydraulic energy supplied is converted into kinetic energy and dissipated in hydraulic losses with the turbine spinning at maximum speed.

The zero speed line is the extreme condition when the turbine is not rotating and all hydraulic energy is throttled in the turbine/nozzle passageways. During normal operation the zero speed line is not encountered.

An important design criteria is the requirement that the turbine must be able to withstand an abrupt speed increase. This will occur when a trip in the electrical system happens. Electrical energy is instantaneously interrupted upon for instance a disconnection from the grid when the electrical circuit opens. The turbine will undergo an almost instantaneous increase in speed from the normal operating speed of approximately 3000 rpm to a speed around 4200 rpm. The operating point transfers to the zero torque line. The unit can rotate under this condition for several minutes without a strict time limit, however the control system will reduce the speed to the minimum operating speed of just over 2000 rpm.

Table 1 lists some typical process requirements for a LNG hydraulic turbine.

Operating Conditions		AGACW	RGCCW	LGHCW	AGACW-L	ELGHC-W
Flow rate	: kg/s	129.22	137.73	117.95	130.47	110.80
Intake capacity	: dm3/s	273.8	286.2	253.0	277.5	240.5
Intake capacity	: m3/hr	985.70	1030.32	910.80	999.00	865.80
Inlet pressure	: bar abs	44.79	43.96	45.27	44.79	44.82
Inlet static head 44.75m :bar		2.07	2.11	2.04	2.06	2.02
Inlet density	: kg/m3	471.8	481.3	466.3	470.1	460.5
Inlet temperature	: C°	-160.2	-160.8	-159.8	-160.1	-162.6
Outlet pressure	: bar abs	2.5	2.5	2.5	2.5	2.5
Δ P of Turbine	: bar	44.36	43.57	44.81	44.35	44.34
Δ H of Turbine	: m	959.95	924.25	981.25	963.25	983.05

Table 1 Typical process requirements.

4.0 DEVELOPMENT STEPS

Although the design of the hydraulic turbine has been based on the well proven submerged motor pumps in LNG service, many new and novel features exist. The requirements for reliability of LNG plants demand proven equipment. In order to permit the application of novel equipment, the confidence level for adequate reliability had to be achieved through a rigorous development programme. This process requires that all steps be formalised by approval of a Development Release under the Shell International Oil Products quality system. These requirements to achieve such Development Release were agreed between the end user and manufacturer in an early stage. The main requirements are:

- Confirmation and listing of proven experience.
- Compliance with Oman LNG Project requirements.
- Mechanical Analysis.
- Fluid Dynamic Analysis.
- Comparison of variable speed versus alternative designs.
- Type testing.

During a formal presentation in February 1997 an in-principle approval to proceed was obtained. Details of further requirements to complete the Development Release programme were agreed upon.
The next paragraphs describe some of the salient steps throughout the development.

4.1 Prototype testing
A prototype single stage turbine was tested to confirm the feasibility. This verified the potential to achieve the required efficiency. Tests were carried out utilising LNG fluid with a 250 kW capacity VSCF unit.

4.2 Computational Fluid Dynamic Analysis

The initial geometry and lay out of the hydraulic components were based on reverse operating LNG pumps. A CFD analysis was carried out to review the flow through a reverse operating pump and its limitations became apparent. Refer figure 5, that illustrates the large energy dissipating vortices due to non optimal flow conditions resulting from a geometry that is based on pump operation. It also illustrated the potential for improvement of efficiency. The CFD analysis was used to optimise the flow patterns by modifying the geometry of the nozzle rings, runners and flow return channels. An improved optimal flow was achieved within the overall dimensions. Refer figure 6, which shows a much improved flow pattern. (3,4)

4.3 Fluid Tests

A main requirement of the development was a verification of the efficiency and performance at an early stage of the manufacturing process. The ultimate verification of the entire unit would consist of a full load full speed string test utilising LNG at the manufacturer works. An early verification of performance was not only required from a project-planning point of view but also for the following reasons. CFD analysis is being applied rather frequently nowadays however verification was required to substantiate the results. The effect of multistage had to be confirmed as only single stages were analysed (CFD) and prototype tested.

It was agreed that a test was performed with a single stage turbine utilising water as a test fluid. A complete stage consisting of inlet nozzle ring, runner and return channel was fabricated at full scale. Refer figure 7. In order to ensure similar fluid flow conditions for LNG fluid the test was performed at EBARA Japan research facilities in accordance with IEC 193,193A International codes for Model Acceptance test of Hydraulic Turbine. The speed for the test was reduced to lower the power requirements and maintain the existing test loop pressure ratings. In order to minimise regret and allow corrections at a late stage, several nozzle rings with slightly varying exit angles, were used in the water test. The added benefit of a full-scale water test was that these nozzle rings were immediately available in case the ultimate test on LNG would require a correction. It also would cover an uncertainty perceived from investigations based on single stages rather than multiple stages.

The outcome of the water tests confirmed the CFD analysis predictions within the accuracy ranges. One major outcome was a reduction of number of stages. The initial design was based on four and five stages for the LNG and Heavy MR turbines respectively. These were now reduced to two and three stages.

4.4 Rotor Dynamic Analysis

A thorough rotor dynamic analysis was required to confirm a robust design and the technical integrity. Two aspects required specific attention: The unit has a variable speed and therefore sufficient separation margins with respect to critical speeds shall exist for all speeds within the operating range. A possible rotor excitation may exist from release of He gas in the last turbine stage. Hence the rotor stability had to be confirmed.

The analysis was based on a complete model that included the casing of the turbine and generator.(5,6) An important feature proved to be the connection of the turbine inlet flange to the headplate of the pressure vessel that contains the entire unit. The stiffness of this joint proved to be a main influence of the rotor dynamic behaviour. In order to confirm the final rotor dynamic results, a modal analysis was performed on the turbine- head plate joint.

Measured data of stiffness values obtained from the as built unit were included in the calculations making these as realistic as possible. It was also confirmed that the gasket thickness and bolt tensioning, within reason, did not largely affect the joint stiffness. Thereby eliminating a source of rotor dynamic problems, that could be introduced during maintenance to the turbine headplate bolting, if incorrect bolting torque were applied.

The analysis that were performed :

- Finite Element Analysis of the hydraulic turbine-generator casing. Refer figure 8.
- Modeling of the integrated system Rotor-Casing. Refer figure 9.
- Critical Speed Analysis of the rotor-bearing-bushing/seal-casing.
- Unbalance Response Analysis of the integrated system. Refer figure 10.
- Stability Analysis of the integrated system.
- Torsional Analysis, Steady State and Transient of the turbine-generator system.

4.5 Electrical Tests
The electrical system related to the hydraulic turbine is depicted in figure 3. The selection of the supplier of this system was carried out as a combined effort between turbine manufacturer, who has overall responsibility for the entire string and the end user. The suppliers of these Variable Speed Constant Frequency systems mainly originate from the wind turbine industry where multiple variable speed turbines supply constant frequency electric power to the grid. Units in the range of 1Mw input have been applied in that industry but are at the top end of the range.

The selected supplier was CEGELEC Industrial Controls using their Delta Power module type GD4000 with Integrated Gate Bipolar Transistor (IGBT). The VSCF generation principle uses a Pulse Width Modulation IGBT scheme to control the variable speed generation (5). The speed control algorithm is a Volt per Hertz calculation scheme. Vary the voltage to vary the frequency of generator excitation in an induction style generator. Hence the excitation frequency defines the speed at which the generator will operate.

Existing technical requirements as normally applied to variable speed electric motor drive systems formed the basis for the turbine electrical system. The VSCF unit features were verified against these requirements and accepted. In particular the existing experience with the GD4000 modules in the power range applicable was reviewed with proven reliability and expected life as the critical important items. Due to the location of the VSCF control cabinets being inside an electrical substation, the cooling system of these units which is important for reliability and component life expectancy, had to be specifically designed to meet the requirements.

All units that have been supplied were subject to a full load test at the VSCF supplier. One complete unit including step-up and step-down transformers was shipped to the turbine manufacturers test facility in the USA and has been used for the full load string test on LNG. The unit proved to operate satisfactory and controlled all aspects in accordance with the specification.

5.0   VERIFICATION AND TEST RESULTS

The factory test and inspection requirements have been specified in detail in the Oman LNG project specification. These requirements apply to all units supplied. The total number of units included one LNG hydraulic turbine and one Heavy MR hydraulic turbine for each LNG train. For each type one complete spare unit was required. This amounts to a total of six turbines and critical components of a VSCF unit.

Apart from the normal requirements specified for submerged LNG pumps the factory acceptance tests included:

- Mechanical running tests to prove technical integrity and confirm vibration levels.
- Performance tests to confirm and validate the required efficiency.
- Endurance runs to prove technical integrity at runaway conditions.
- Strip down and component inspection of the unit after to completed tests.
- Tests will utilise the job VSCF system and voltage transformers under full load conditions.
- All tests to be performed utilising liquid LNG, flow and head values to be identical as applicable for the site.

The acceptance tests proved to be successful and a synopsis of some test results are included in table 2 and figure 11. The required efficiency values were achieved.

Figure 12 shows a photograph of the manufacturers test stand and the Oman LNG hydraulic turbine after completion of the Full Load String Test.

Operating Case		Capacity (m³/hr)	Head (m)	Power (kW)	Efficiency (%)
AGACW	Predicted	986.22	898.4	898.5	79.3
	1-GT-1420	986.2	894.0	924.5	81.5
	2-GT-1420	986.2	902.2	935.1	81.7
	GT-1420 S	986.2	902.7	933.7	81.6
	Average	986.2	899.6	931.1	81.6
RGCCW	Predicted	1030.32	863.82	898.5	77.4
	1-GT-1420	1030.3	866.9	908.7	77.6
	2-GT-1420	1030.3	864.5	914.0	78.3
	GT-1420 S	1030.3	869.8	916.3	78.0
	Average	1030.3	867.1	921.7	78.0
LGHCW	Predicted	910.80	923.62	810.9	76.3
	1-GT-1420	910.8	930.9	869.8	81.3
	2-GT-1420	910.8	922.5	865.5	81.5
	GT-1420 S	910.8	929.6	869.1	81.2
	Average	910.8	927.7	868.1	81.3

Table 2 Typical test results.

6.0 CONCLUSION

The manufacturing and testing of variable speed hydraulic turbines for LNG plants has been successfully completed in accordance with the agreed requirements and within the agreed time frame. The units have been shipped to Sur in Oman for installation in the LNG complex.

LNG plants demand high reliability in order to meet long term contractual delivery schemes. Credible reliability figures are mainly based on proven equipment with an actual operating history. Novel and new equipment can only be applied in LNG plants if the design is an extrapolation of existing design and novel aspects are carefully reviewed and analysed. The end user quality system defines a Development Release as a formal process to evaluate this.The development requirements and steps need to be defined from the beginning and to be agreed between manufacturer and end user. Review of the results as these become available during the development process by all parties involved, is necessary. Adjustment of the programme requirements may prove to be necessary in view of intermediate results, requiring a flexible approach and open dialog.

7.0 EFERENCES

1   Kimmel H.E.; "Variable Speed Turbine Generators in LNG Liquefaction Plants", Proceedings of GASTECH '96,Vienna, Austria, December 1996.
2   Habets G.L.G.M., Kimmel H.E.; " Economics of Cryogenic Turbine Expanders", The International Journal of Hydrocarbon Engineering, December 1998.
3   Baines N.C.,Oliphant K.N., " Cryogenic expander turbine", ETI Concepts Technical Memorandum 504 , April 1997.
4   Baines N.C., Oliphant K.N.,Kimmel H.E.,Habets G.L.G.M.,"CFD analysis and test of a fluid machine operating as a pump and turbine", Proceedings of IMechE Seminar, Publication S546/008/98, October 1998.
5   Smalley A.J.,Habets G.L.G.M.,Kimmel H.E.,Hollingsworth J.L.,"Rotor Dynamic Analysis of a Submerged Turbine Generator Driven by Liquefied Natural Gas" Proceedings 5th International Conference on Rotor Dynamics September 1998,ISBN 3-528-03870-5
6   Habets G.L.G.M., et al. "Rotor Dynamic Analysis of a Cryogenic Submerged Turbine Generator" Proceedings of the 19th Southeastern Conference on Theoretical and Applied Mechanics ,Boca Raton , Florida , April 1998, ISBN 0-9663841-0-05.

C556/008 © IMechE 1999

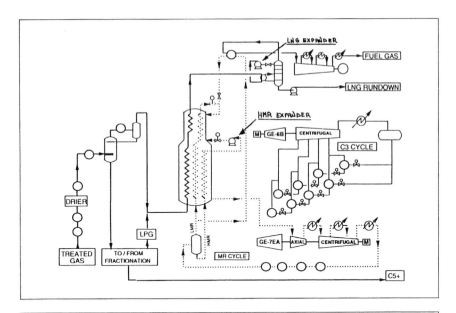

Figure 1 : Overall Plant process scheme - typical

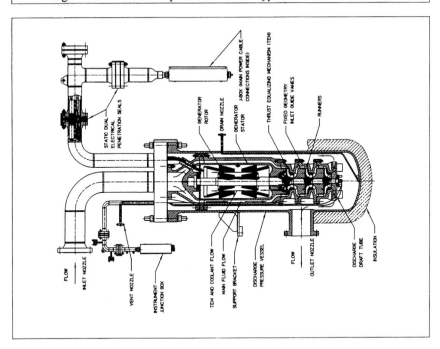

Figure 2 : Turbine cross section – Type 10 TG – 153.

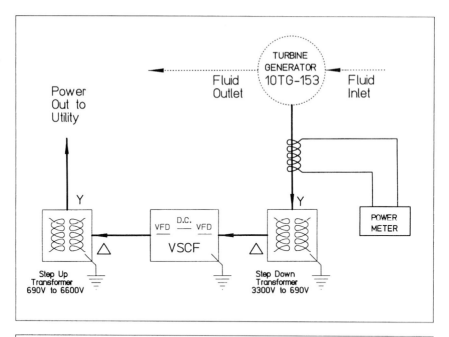

Figure 3 : Electrical system – overview VSCF.

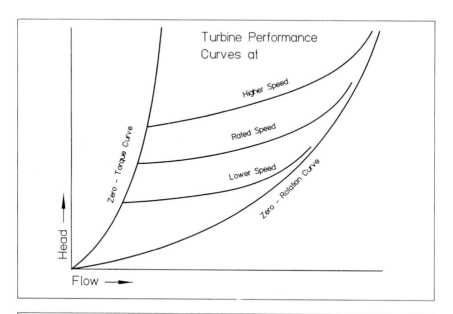

Figure 4 : Typical performance curve for variable speed radial hydraulic turbines.

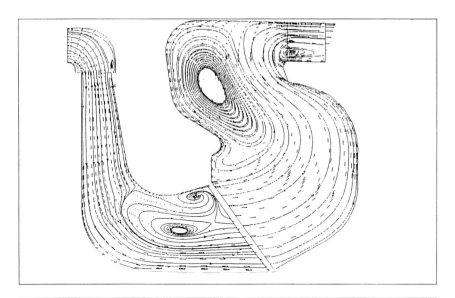

Figure 5 : Initial geometry – flow pattern results from CFD for runner and return channel.

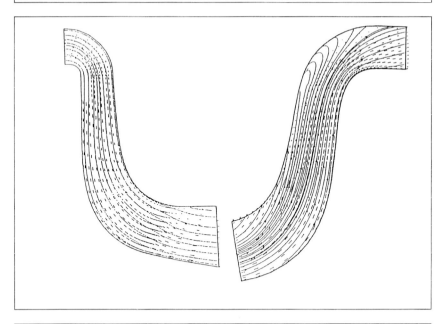

Figure 6 : Modified design – flow pattern results from CFD for runner and return channel.

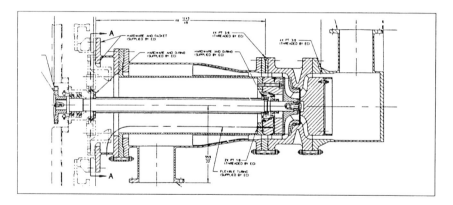

Figure 7 : Full scale single turbine stage in a water test rig – test rig outline.

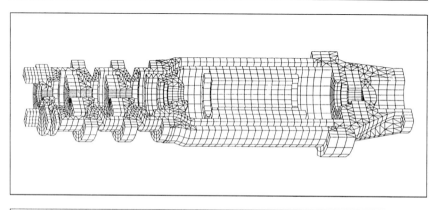

Figure 8 : Finite Element Model of a turbine-generator casing.

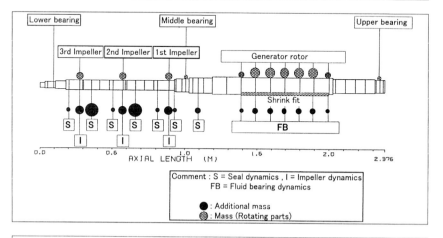

Figure 9 : Rotor – Casing lumped mass model.

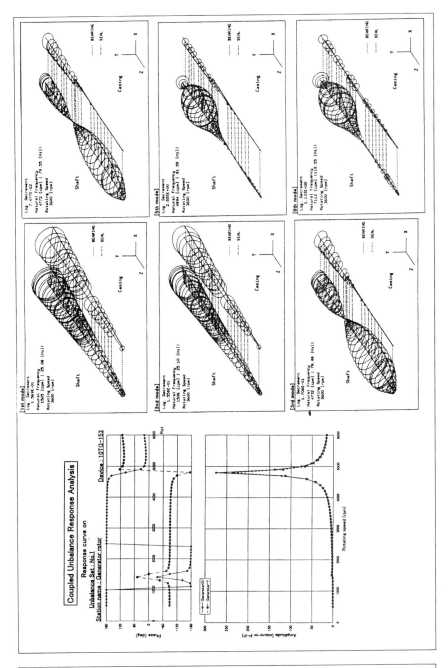

Figure 10 :   Response analysis results for shaft –bearing-casing.
        10 a :   Unbalance response – Generator mid span.
        10 b,c : Stability analysis – Mode shapes rotor and casing.

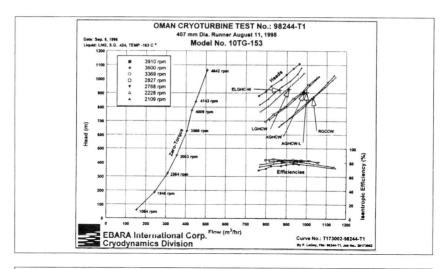

Figure 11: Performance map of Oman LNG Hydraulic turbine – typical results.

Figure 12 : Oman LNG unit after completion of Full Load String Test.

C556/011/99

Free Floating Piston™ technology – a new development to increase the life of reciprocating compressor wear bands

R J M SCHUTTE
Thomassen International BV, Rheden, The Netherlands

SYNOPSIS

For large horizontally opposed compressors, changing process and environmental requirements are dictating the increased use of non-lubricated cylinders. Improving the sometimes unpredictable life of piston wear bands remains an elusive goal.

Thomassen, a manufacturer of such machines, wished to achieve a consistent and substantial improvement in wear band life. To achieve this goal, a radical approach was attempted.

A feasibility study to evaluate alternative bearing forms for pistons, left magnetic and gas bearings as main candidates, The gas bearing proved ultimately to be the more attractive option. A calculation model was developed, and verified on a static test rig. After successful static trials the design was incorporated into a full size, heavy duty test compressor. Inspection after 2000 hours of operation revealed only traces of bedding-in wear on the wear bands.

The Free Floating Piston™ technology may be used in almost all compressor applications, with all types of process gas, with virtually no wear of wear bands. It can easily be retrofitted to existing machines by virtue of its self contained design.At present a FFP™ technology equipped hydrogen compressor is in full time operation in the Netherlands.

Financial assistance for this final phase was provided by the Netherlands Ministry of Economic Affairs. The design has been filed for patents in a number of countries around the world.

This paper describes:
- A summary of the gas bearing technology.
- Application in a reciprocating compressor.
- Design.
- Testing and test results.
- Research path.

INTRODUCTION

Thomassen Compression Systems has manufactured reciprocating compressors since 1906, mainly of the horizontal balanced opposed type. The crankshaft, connecting rod and crosshead mechanism, in principle, require lubrication. The piston may be lubricated or non-lubricated. depending on process gas or environmental constraints.

As compressors may be critical components in production plants, these are often equipped with one operational and one spare compressor, to prevent production losses during maintenance or calamities. To achieve low operation and maintenance costs, a reciprocating compressor is required to operate uninterruptedly for as long as possible. A target of three years of continuous operation is given in the main industry standard, API 618.

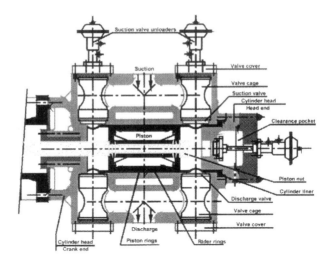

Cylinder

Figure 1: Sectional drawing of cylinder

The piston is fitted with piston rings and wear bands, or rider rings. The piston rings act as a seal between the high and low pressure side of the cylinder. Because of its weight, the piston is supported by rider rings which keep the piston free of the cylinder wall. It is usual to make both piston rings and rider rings of a synthetic material such as a PTFE composite. In non-lubricated service, wear, especially of the rider rings, tends to be unpredictable. Certain combinations of process and gas conditions can increase the wear rate so drastically that running time can be reduced to as little as a few hunderd hours. The economic pressure to achieve an uninterrupted running time of three years or more has made improvement in wear rate the subject of research over a long period of time.

The usual approach has been to attempt improvement of the wear properties of rider ring and liner materials. As this has only produced marginal results, Thomassen decided to take a more fundamental approach to achieve improvement. The resulting research project has led to a design called Free Floating Piston™ technology.

FEASIBILITY STUDY

The start of the research project was the execution of a feasibility study to examine the fundamental options. This study was performed as a graduation project, in which the following three alternatives were assessed:
- a permanent magnet bearing;
- an electromagnetic bearing;
- a gas bearing.

The Permanent magnet bearing
The Permanent magnet bearing operates on the principle of like pole repulsion. Permanent magnets are passive elements requiring no energy supply to obtain load carrying capacity. They are used for radial bearings in small electromechanical components such as motors and actuators.

The Electromagnetic bearing
The principle is similar to the permanent magnet bearing. It differs from the permanent magnet bearing in that it is both active and pro-active. It is active as an energy supply is required to obtain functionality and pro-active as displacement between poles can be regulated.

The Gas bearing
Gas lubrication may be classified as a branch of viscous compressible fluid machanics. More than a century ago Kingsbury wrote a publication on the feasibility of externally pressurized bearings. Practical applications were not possible until the second half of this century due to the lack of adequate manufacturing processes. Calculation of gas bearings behaviour is complex due to influence of fluid compressibility. The dentist's drill and the railway workshop train transport platform are present day examples of gas bearing applications.

For the use of one of these three bearing types in a reciprocating compressor, the location of the bearing proved to be a major constraint in the choice of bearing type.

Three locations were considered feasible:

– directly on the piston.
– indirectly: on the piston rod.
– indirectly, duplex: on piston rod and tail rod, that is, straddling the piston.

To enable selection of a bearing type, it was necessary to formulate evaluation criteria, for which the following were chosen:

– Elimination of unpredictable rider ring wear.
– Suitable for retrofitting to an existing compressor at an acceptable price.
– High reliability.
– "Fail Safe" properties.

Evaluation of the options yielded the following conclusions:

– The electromagnetic bearing is an expensive design requiring major design changes of the cylinder, cylinder compartment, piston and piston rod.
– Permanent magnets are feasible in theory, but are physically impractical due to the alignment problems associated with the requirement for very small pole pieces. A further disadvantage is that any magnetic particles in the process gas would contaminate the magnets.
– At this stage the gas bearing appeared to be the most attractive option.
– The preferred location of the bearing proved to be directly on the piston. Analysis of a bearing located on the piston rod, revealed that a piston could have a vertical displacement of nearly1 mm due to piston rod flexibility. An extra bearing on a tail-rod would reduce this to the order of magnitude of 0.3 mm. To meet the requirements for a stable and controllable bearing,support directly at the piston proved to be necessary.

GAS BEARING DESIGN

Gas bearing designs may be divided into two categories, aerodynamic and aerostatic. In an aerodynamic bearing the fluid film pressure distribution and resulting load carrying capacity are generated by the velocity difference between the interactive running surfaces. In an aerostatic or externally pressurized bearing, the fluid film and load carrying capacity are derived from an external source of gas under pressure. As the velocity difference required for an aerodynamic bearing cannot be generated by reciprocating movement, an external gas supply under pressure is used for the FFP™ technology.

Nominally, zero contact between running surfaces is the interface condition when the force generated by the distributed pressure equals the externally applied load. The load characteristic of the primitive aerostatic bearing is such that a disturbance in equilibrium means surface contact with a load increase, and in a theoretically infinite clearance if the load decreases. This is an indeterminate bearing characteristic as the magnitude of the external load does not define a unique shaft position. To be stable, the bearing must have a suitable stiffness characteristic. The desired stability may be introduced into the bearing characteristic by the addition of flow resistance in the gas supply. An additional improvement is also possible. The area integral of the distributed pressure between the running surfaces determines the total bearing load carrying capacity. To increase the load carrying capacity, it is possible to design a surface profile to influence the pressure distribution. This may be employed to enable a lower external pressure to yield the required load carrying capacity.

The principal difference between gas and liquid bearings is that gas is a compressible medium. As a result, the gas flow cannot be calculated as a volume but must be calculated as mass flow, taking into account gas density and compressibility. Another difference is that the pressure equilibrium cannot be calculated directly, as no explicit analytical equation can be formulated. The number of parameters to be calculated exceeds the number of equations, making this approach impossible. An iterative procedure, using initially assumed values, ultimately results in unique solutions for load equilibrium conditions.

CALCULATION MODEL AND DESIGN

Figure 2: Free Floating Piston™ technology

The chosen configuration for Free Floating Piston™ technology is that the gas supply for external pressurization enters the hollow piston through valves in the piston faces during the compression cycle when the cylinder pressure exceeds the internal pressure of the piston. (figure 2). The cylinder is usually double acting so that the pressure charge is twice per revolution, but may also operate successfully in a single acting mode. The piston acts as a gas reservoir. The valves are ordinary small compressor valves, acting on differential pressure. The gas for pressurization exits the piston through holes in the bottom section of the rider rings which are a shrink fit on the piston. To obtain a reasonably constant pressure environment for the rider rings, the piston rings are located outboard of the rider rings. Pressurization gas flow through the gas bearing / rider ring is small in comparison to piston ring blow-by. Spent pressurization gas also escapes as blow-by. Pressurization gas consumption is so small that it does not affect the compressor capacity to a significant degree. The outlined Free Floating Piston™ technology has been filed for patent in a number of countries.

The design of FFP™ technology gas bearing starts with determining the location of the rider rings on the piston within applicable geometric constraints. Rider rings remain a PTFE based material for their self lubricating properties, serving for starts, stops and failure mode operation, during which they function simply as non-lubricated rider rings. Starts and stops are in fact beneficial as the resulting running-in leads to improved surface conformance and an optimum bearing performance. Once the rider ring location has been established, the load carrying capacity can be calculated. A finite element method is used to calculate the load division between the rider rings with the gas bearings modelled as stiff springs.

As the peak value of the pressure distribution over the running surface was unknown, measurements were performed to assist in the formulation of the calculation model. With bearing load as input, the calculation model is able to reveal the peak pressure. This is required input for the equations which must then be solved, which yield properties of the gas bearing such as flow film thickness and intermediate equilibrium pressures.

Once this has been completed, the bearing stiffness may be calculated, the outcome providing the new input for recalculation of the design. Thereafter, the actual load distribution may be calculated.

TESTING AND RESULTS

Figure 3: Static test rig

Static test rig

A static test rig was built for two main reasons. First of all, the application of the gas bearing integrated in a reciprocating compressor piston was, to our knowledge, untried. Simply scaling up other gas bearing research results would not have been satisfactory, as dimensions and ratios between determinant parameters differ tremendously. A second reason was that pressure distributions for the dimensions involved had to be measured under various conditions, to enable the development of a calculation model.

The test rig consisted of a steel pipe as a surrogate cylinder liner and a piston composed of a sandwich of several steel rings and two covers. A rod was fitted through the piston to be able to attach weights for increasing in the bearing load. Although the piston rings would normally be manufactured from a PTFE composite material, the test rig piston rings were made of steel for optimum surface shape and finish.

The first series of tests were executed to measure and map the developed pressure distribution between the running surfaces.

Figure 4 shows an example of the projected side view of a measured pressure distribution. The figure shows three maximum pressure peaks along the circumferential axis due to the three gas supply holes in the rider ring. The peak shapes are also a result of the chosen ring profile. Various load conditions, bearing profiles and rider ring shapes were tested to develop an analytical formula to permit the calculation of the pressure distribution. The pressure distribution characteristic proved to be primarily a function of load; discharge pressure variation having only a minor effect. The established pressure distribution characteristic has therefore, a general applicability to pistons.

The second test series focussed on film thickness and mass flow through the bearing. General equations for similar flow conditions were available from subject literature, so that testing was focussed mainly on the shape, dimensions and flow resistance factors of the flow nozzle, as well as film entrance resistance.

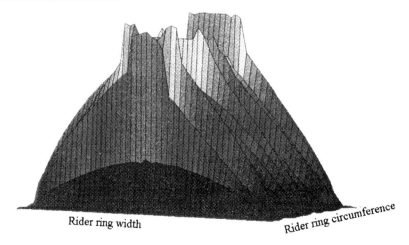

Rider ring width   Rider ring circumference

Figure 4: Side view pressure distribution

The critical factor in the design of a gas bearing for this application is its stability. The term "pneumatic hammer" is to be found in literature to describe a condition where a critical flow velocity causes resonance of the piston on the gas film. Criteria were developed to be able to avoid this mode of instability.

APPLICATION IN A RECIPROCATING COMPRESSOR

Dynamic testing

After a succesful test series on the static test rig, Thomassen's heavy duty test compressor was equipped with Free Floating Piston™ technology (figure 5). As the piston dimensions in the static test rig were almost equal to those of the compressor, the design could virtually be copied. Initially, an external gas supply to the piston and gas bearing connected by tubing, was used. This meant that the compressor was made directly comparable to the static test rig, but that test results could be verified under more realistic conditions: that of actual machine operation.

Figure 5: Dynamic test stand

After verification of these test results, the external gas supply was disconnected. Testing was continued with the valves located in the piston to provide auto-pressurization for the bearing gas supply. The Free Floating Piston™ technology functioned as calculated and a life behaviour test sequence was performed over a period of more than 2000 hours with about 150 compressor starts and stops.

Field application

As testing and step by step development indicated that concept of FFP™ technology was successful, a duration test in a field application was prepared. As most reciprocating refinery compressors do not compress air but other gases with different properties, the calculation model had to be expanded to include other gases. It became possible to install Free Floating Piston™ technology in a hydrogen compressor in commercial operation. It had higher pressures and power requirements than the Thomassen test compressor, but as the dimensions

C556/011 © IMechE 1999

were very similar to the Thomassen test compressor it represented a final step in the systematic development process.

A condition monitoring system is connected to the compressor with an on-line telephone connection to Thomassen. Operating conditions, performance and wear checks are performed regularly. The performance to date exceeds expectation, and the wear of the rider rings, the basic reason for the whole development project is virtually non-existent. While this was predicted its actual occurance is novel when one is used to wear as a fact of life. At the time of writing this hydrogen compressor had some 4000 hours of operation without detectable rider ring wear. A sample of data shown below gives an indication of the monitored parameters.

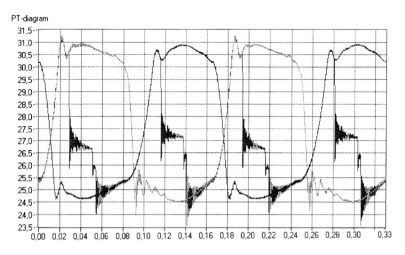

Figure 6: Pressure versus time

Figur 6 shows a pressure versus time measurement taken from the hydrogen compressor during normal operation. Three pressure transducers have been fitted to the cylinder. Two measure the pressure in the clearance volumes, for the plotting of pressure/volume diagrams and the calculation of power requirements. The third pressure transducer is located midstroke in the cylinder. The pressure it measures reveals, amongst others data, the ambient pressure at the rider rings. This is an essential parameter to assess the operational behaviour of Free Floating Piston™ technology.

Financial assistance for the field application phase was provided by the Netherlands Ministry of Economic Affairs.

CONCLUSIONS

Free Floating Piston™ technology is a significant advance in compressor design. Wear is limited to start and stop conditions and is therefore practically absent. Maintenance requirements for piston rings and stuffing box elements have diminished, as vertical piston displacement due to wear is zero. By eliminating rider ring wear, other components become the limiting factor for maintenance intervals. Compressor reliability and availability are improved; production losses and maintenance costs will decrease.

Cylinder lubrication systems become unnecessary, resulting in fewer compressor auxiliary systems. Process gas contamination and valve fouling belong to the past. Measures to meet environmental regulations for oil pollution become superfluous. FFP™ technology is simple to install in both new and existing compressors as it is a very minor change in the overall compressor design, having no impact on the rest of the compressor.

To conclude, Free Floating Piston™ technology has proved to be a simple and elegant solution to an old and often nettlesome problem. It is attractive due to its self contained nature and its ease of installation and use in both new and existing compressors without further modification. It has proved to be both very effective and reliable in practice.

C556/011 © IMechE 1999

C556/021/99

Magnetic bearings – an alternate for turbomachinery

Y DESTOMBES
Société de Mécanique Magnétique-S2M, Saint Marcel, France

SYNOPSIS

Considered more and more often as a valuable alternate to oil lubricated bearings, magnetic bearings have to date experienced successful operation in a significant number of industrial turbo-machines. This proven technology offers several obvious advantages in the oil and gas industry, mainly because of the elimination of the lube oil unit, the high reliability of the system and the easy arrangement and operation of the machines which integrate these bearings.

1. INTRODUCTION

The first manufacturer to experience magnetic bearings on an industrial machine was Ingersoll Rand in 1980 with a full 5 axis magnetic suspension. This test machine, working with air and nitrogen, had a rating of 4MW with a rotation speed of 13,000rpm. The first industrial compressor with magnetic bearings was a Ingersoll Rand pipeline compressor rated at 12 MW and 5,250 rpm. This machine was commissioned in 1985 for a pipeline booster station in Alberta/Canada. Since that time, more than 200 turbo-machines (mainly compressors and turbo-expanders) supported by magnetic bearings have been put in operation in many different applications in the oil and gas industry and hydrocarbon processing.

2. WORKING PRINCIPLE

2.1 Mechanical

As for any bearing, the active magnetic bearing consists of a fixed stator which supports a rotating shaft with the minimal friction and wear. Between the rotor and stator parts the carrying force typically provided by an oil film is replaced by magnetic forces which support the rotor inside of the machine. Since these forces are applied across a small clearance gap between the rotor and stator, there is no mechanical contact which essentially eliminates friction inside the bearing!

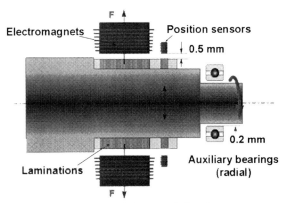

Figure 1 : Radial magnetic bearing.

The magnetic field which creates the force is generated by stator electromagnets located around the shaft. This field passes through the rotor, which is equipped with ferromagnetic laminations in order to minimise the Eddy current losses in the rotor during rotation. The laminations also provide a high permeability to the magnetism which helps to concentrate the magnetic flux at the bearing and minimises flux leakage. The fields created are fixed in the main two perpendicular axes of freedom - V & W - oriented at +/- 45° from the vertical direction. The position of the rotor inside of the stator is detected by inductive sensors located close to the electromagnets of the bearing also oriented in V & W axes.

2.2 Technology
The technology of a radial magnetic bearing is very close to that of an induction motor with electromagnets on the stator (outside) and laminations on the rotor (inside). The stator is made of laminations with slots and windings used to generate the magnetic flux that is created under the poles. This arrangement also includes the inductive position sensors for which the technology is very similar to that of the bearings. The radial clearance of the magnetic bearings is adapted from 0.3 mm to about 1 mm according to the diameter of the bearing and the application, but is generally set to 0.5 mm for medium size turbo-machines. Axial control of the machines is performed by means of a double acting thrust bearing, also of the electromagnetic type. The stationary ring shaped electromagnets, as shown in Figure 2, are located on both sides of the thrust disk which is mounted on the shaft.

The thrust disk is a solid piece of steel and does not have to be laminated like the radial rotor because there are no Eddy current losses since each point of the disk remains under the same magnetic pole. The associated axial sensor is not linked to the thrust bearing stator and is most of the time combined with the radial sensor.

C556/021 © IMechE 1999

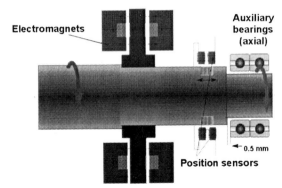

Figure 2 : Magnetic axial thrust bearing.

In order to protect the electromechanical parts in case of bearing overload or electronics failure, a set of auxiliary bearings is used to limit both radial and axial motions. The auxiliary bearings are fixed at the stator, with a smaller clearance (0.2 to 0.5 mm) to the rotor. These bearings are designed for five de-energised landings.

2.3 Feed-back control system.

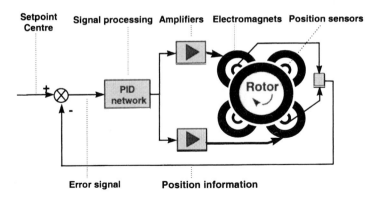

Figure 3 : Controller close loop function.

The position sensors are arranged in a bridge configuration which is driven by a high frequency signal to provide a linear output signal indicating the rotor location. This signal is compared with the reference set point (generally zero when a centred rotor is desired). Any deviation from the centre will result in an error signal which is treated in a PID network (P = Proportional (position), I = Integral (stiffness), D = Differential (damping)) and amplified in the power amplifiers. The amplifiers adjust the current to the electromagnets in order to hold the shaft in the centred position.

As a result, the possible displacements of the rotor initiated by external load variations are cancelled by counteracting electromagnetic forces generated by the stator. The modulation of these forces is indeed a modulation of the currents or, better said of the flux according to the actual location of the rotor detected by the position sensors. A full rotor levitation consisting of five degrees of freedom or five axes requires ten electromagnets arranged with two on each side of the axis.

An additional electronic device allows a suppression of unbalance force transmission from the rotor to the stator for and only for the synchronous signal (synchronous to the speed). This Automatic Balancing System (ABS) allows the shaft to rotate about its inertial centre resulting in a vibration free operation of the stator and thus much less force is transferred to the machine frame and foundations.

The position and control information is available in the form of electrical signals in the electronic cabinet so that an extensive monitoring of bearing, machine and process parameters can be achieved thanks to the bearing reactions.

Bearing function (i.e. levitation) is effective upon powering the electronics and does not require any dynamic effect from the speed or « warm up » sequence.

3. APPLICATIONS TO OIL AND GAS COMPRESSORS.

One of the merits of this bearing technology is to replace complicated and expensive lube oil systems (in particular for API machines) by simple and robust industrial electronics which require much less attention and maintenance.

3.1. Nevertheless, some configurations may not entirely get rid of the oil systems :

3.1.1. Centrifugal compressors with low speed drive and step up gear:
In this configuration the use of magnetic bearings eliminates the oil in the compressor itself, but does not simplify the skid, because a lube oil unit is necessary for the gear. The justification for the choice of a magnetic bearing in this case is to increase the machine reliability. Today, six compressors of this kind are in operation with Thermodyn for export gas and booster compressors in Europe, Elliott, Dresser-Rand, and IMO-Delaval for refinery application in North America.

3.1.2. Centrifugal compressors driven by a conventional oil lubricated power turbine.
This type of compressor configuration is mainly used in pipeline applications for beam type or overhung impeller machine designs. Magnetic bearings offer their advantages in reduced maintenance and remote and centralised control. The evolution of independent synthetic oil systems, as substitute to mineral oil lubrication for the bearings of the drive (power turbine), justifies even more the use of magnetic bearings for such compressors. The reliability values of these machines achieve and surpass equivalent oil bearing compressor reliability in similar applications, reaching 99.9% over several years. More than 30 compressors of this type are in operation in Canada.

   C556/021 © IMechE 1999

Picture A : Centrifugal compressor for gas pipeline. (by courtesy of Nova GTL)

3.2. Most of the references to date apply with total elimination of the lube oil system.

3.2.1. Centrifugal compressor driven by a steam turbine (both on magnetic bearings).

The feasibility of magnetic bearings for steam turbine has been first proven by the construction of a 3 MW, 15,000 rpm test steam turbine with MHI (Mitsubishi Heavy Industries), using magnetic bearings. Early in 1997, a propane refrigeration compressor string manufactured by Elliott has been commissioned. It consists of a multistage steam turbine directly driving a centrifugal compressor. Both machine bodies are equipped with magnetic bearings. Though the process commissioning has been longer than expected, the bearings have demonstrated their capabilities both on the compressor and the turbine and are in satisfactory operation conditions.

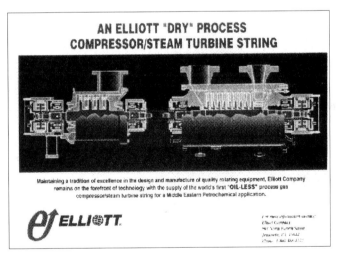

Picture B : « Dry string » with steam turbine driven centrifugal compressor for MTBE process. (by courtesy of ELLIOTT)

3.2.2 "Low speed" centrifugal compressor.

Some processes are using compressors at speeds not exceeding 3,000 or 3,600 rpm. In this configuration of direct « synchronous » drive, there is little advantage to use magnetic bearings for conventional duty. But in the particular case of polyethylene compressors such as the overhung ones designed by Demag Delaval, the combination of dry gas seals and magnetic bearings fully eliminates the risks of oil contamination to the process. Another important advantage in this application is the ability of the magnetic bearing to handle the high level of unbalance that is caused by polymer build up on the impeller. The first machine of this type is in operation in France since 1991. The next ones in Finland since 1993, in Ukraine in 1996, followed by 6 other compressors in India, again for polyethylene process.

3.2.3. « High speed » centrifugal compressor driven by a directly coupled high speed electric motor.

Solid state variable frequency drive for high speed, high power induction motors is a recent technology which appeared on the market at the end of the 80's. This is an ideal application because the gearbox is eliminated and the complete string can be dry by applying magnetic bearings to both the motor and compressor. The first industrial centrifugal compressor directly driven by a high speed motor, both compressor and motor bodies with magnetic bearings, is in operation since June 1993 for gas storage at 150 bars in Northern Germany. The Sulzer six stage compressor and the ACEC motor are running at 20.000 rpm, 2 MW totally "oil-free". The digital controller and the consequent remote control capabilities have demonstrated their high availability rate. After 6 seasons of operation, the user confirms it is the best of his 13 compressors, without experiencing any failure on the bearing systems.

Identical concept is presently under works tests but with 2 compressor casings, one at each end of the motor, for gas re-injection at 320 bars. Motor power is 8 MW at 14,700 rpm.

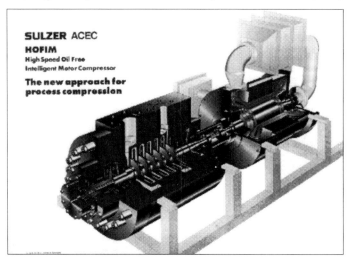

Picture C : Sulzer centrifugal compressor driven by an Alstom induction motor.
(by courtesy of Sulzer TZ & Alstom/ACEC Energie)

C556/021 © IMechE 1999

3.2.4. *MOtorised PIpeline COmpressor (MOPICO)*

The combination of the two above mentioned innovative technologies (High speed electrical drive and Magnetic bearings) has been used by ACEC and Sulzer to design a new machine called the Mopico (MOtorised PIpeline COmpressor). The concept is a single shaft high speed motor supported by magnetic bearings with one overhung wheel at each shaft end. All the rotating elements are inside of one single closed housing and are in direct contact with the process gas. The simplicity and the compactness of this machine makes it easier to install. The clean « oil free » concept is a major issue for the protection of the environment. The first compressor of that type (6 MW, 10,000 rpm) is in operation in Alabama/USA since 1991. There are three additional ones close to Baltimore/ USA since 1994 and three others in a station in the UK started mid '98. Four new units should also be commissioned in 1999 in two compressor stations (one in the UK and one in Quebec/Canada).

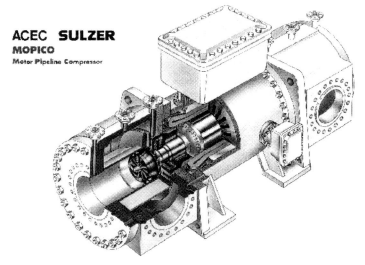

Picture D : Sketch of motor/bearings arrangement in the MOPICO.
(by courtesy of ACEC Energie and SULZER TZ)

4. APPLICATIONS FOR GAS EXPANDERS.

The expander design with a single shaft and one wheel (one compressor and one expander wheel) on each side is the ideal application for this dry bearing type which replaces the whole lube oil system by an electronic device of limited size (2,000 mm x 1,000 mm x 600 mm) and very low maintenance.

The expanders' duty is to cool down the gas by expansion as a more efficient alternate to JT valves in gas cooling processes such as air separation, ethylene production and natural gas treatment. The bearing system is located between the two wheels fixed at each end of the shaft.

The compressor side is warmer with temperatures as high as +80°C, the expander side (i.e. turbine) is cold with temperatures that can go down to -190°C average for air separation processes.

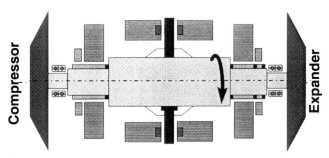

Picture E : Sketch of magnetic bearings arrangement in an expander.

With conventional oil bearing system, process contamination by the lube oil (or vice versa!) together with large temperature gradients require special attention to the sealing design. The use of magnetic bearings eliminates the possible ingress of oil inside of the core of the machine. In applications where the bearings must operate directly in a process gas, a special protection is used to address the problem of chemical aggressivity. The machine is simplified both inside and outside by the elimination of the oil package. The unbalance compensation which equips most of the magnetic bearing controllers for high speed machines is used here for a stress and vibration free function . In addition, the magnetic bearing which works independently from the direction of rotation, accepts reverse rotations that might occur incidentally. Some of these applications use the axial load indication and monitoring from the magnetic bearings (variation of the current in the electromagnets) as a way to balance the thrust inside the machine .

To date, close to 100 machines of that type manufactured by Atlas Copco Rotoflow, BOC Cryostar or Mafi-Trench are in operation all over the world and demonstrate the good reliability of this bearing design.

5. CONCLUSION.

The feasibility and advantages for the user of magnetic bearings in most of the oil and gas processes have been clearly demonstrated by many installations with extensive operating experience. This fact is witnessed by users who have fixed their choice on this kind of equipment.

Although it will never be (and doesn't intend to be) a substitute to conventional bearings in all applications, the economical viability is now proven, mainly when the lube oil system can be completely eliminated.

C556/021 © IMechE 1999

REFERENCES :

•« Magnetic bearing improvement program at Nova/Paul S. Alves, Barry M. Alavi/ ASME Birmingham-UK/ June 1996.

•« Turboexpanders with Magnetic Bearings for Offshore Applications»/ Compressor'Tech/ March, April 1998.

C556/025/99

Optimizing stator vane setting in multistage axial flow compressors

R L ELDER, J SUN, and **P R WIESE**
DTEM/School of Mechanical Engineering, Cranfield University, UK

Further improvement in axial compressor performance is most likely to be achieved through advances in intelligent control, particularly in control of individual vanes and blades. Studies have shown that if the geometry of the individual vanes and blades could be intelligently controlled, the need for a significant surge margin can be reduced, and the performance of a compressor could be significantly improved. The authors have been working on the concept of the Intelligent Geometry Compressor (or IGC), and the recent advances in "agent" control technology appear to make feasible the control of very many variables (such as individual vane geometry). This paper takes the first step along this path, by considering the aerodynamic optimisation of the stator vane angles alone, using stage-stacking methods combined with an optimisation routine. The basic problem is how to optimally set the variable guide vane settings, for any particular aerodynamic condition. A complimentary paper on the optimisation of the control logic will be published shortly. It is appreciated that many compressors do not incorporate sufficient variability, but this analysis could be used to demonstrate the advantages of control of these variables, the cost effectiveness depending on the application.

The paper is presented in four parts: modelling the effect of stagger setting on individual stage performance, stage stacking to provide overall performance including surge point prediction, stagger setting optimisation, and numerical examples. Results from studies on a high speed compressor have demonstrated that variable stagger is a powerful method to rematch stages which can be used to improve desired overall performance. Parametric studies on the optimisation algorithm have also been conducted where it demonstrated numerical stability and fast convergence.

Keywords: variable stagger, compressor vane setting, compressor instabilities, compressor surge, off-design operation, axial flow compressors, compressor optimisation.

NOMENCLATURE

$A(x)$ cross-section area

C	matrix
C_p	specific heat at constant pressure
C_v	specific heat at constant volume
d	averaged distance from centroid to each vertex of polyhedron
f	general performance function
f_d	rotor deviation factor
g	general constraint function
N	compressor design speed
n	number of independent variables
n_c	number of constraints
P	penalty function and total pressure
P_r	overall pressure ratio
P_s	static pressure
R_{2t}	rotor tip radius at inlet
R_{3t}	rotor tip radius at outlet
R_h	rotor hub radius
R_{rms}	root square mean radius
R_t	rotor tip radius
r_k	penalty factor
T	total temperature
T_s	static temperature
U_{2t}	tip blade speed at inlet
U_{3t}	tip blade speed at outlet
V	absolute flow velocity
V_a	axial component of V
V_2^1	relative velocity at rotor inlet
V_3^1	relative velocity at rotor outlet
W	flow rate
W_{sg}	surge mass flow rate, kg/s
X_i	non-dimensional independent variables
$\vec{X}_{dyn}$	perturbation vector of dynamic variables
$\vec{x}$	independent variable vector
β	stagger angle, degree
η	adiabatic efficiency
ρ	density
ϕ	flow coefficient of stage
ψ	pressure rise coefficient
ΔPr	increment of Pr
ΔW_{sg}	increment of W_{sg}, kg/s
ΔX	non-dimensional step length of iteration
$\Delta\beta$	variation of stagger angle from design setting, degree
$\Delta\beta l$	lower bound of $\Delta\beta$, degree
$\Delta\beta u$	upper bound of $\Delta\beta$, degree
$\Delta\eta$	increment of η
$\Delta\delta$	change of rotor deviation
subscript:	

1   stage inlet
2   rotor inlet
3   rotor outlet
4   stage outlet
d   value for nominal (design) conditions
f   at fractional speeds
ref   reference (operating) point

1. INTRODUCTION

Multistage axial flow compressors with a wide operating flow range and high efficiency are required for many industrial applications. The operation of such compressors at lower flow rates is limited by instabilities which cause a full breakdown of the flow which is termed surge. These instabilities, which are caused by high incidence and subsequent stalling of stages, occur due to different phenomena at part and full speed operation [1, 2]. The problem at part speed operation is that the front stages are often heavily stalled and rear stages choked whereas at high speeds, the front stages are operating close to choke and the rear stages tend to be stalled thus causing surge. Optimisation of the design to high speed conditions can often provide part speed problems and to achieve the acceptable performance at part speed operation, variable inlet guiding vanes (IGVs) and the front stage variable stator vane settings, fig. 1 [3], are sometimes used to modify the flow angles to avoid stage stall and subsequent compressor surge [4, 5]. To date, such variable settings follow some schedule established by analysis and experiment whereas this paper proposes a methodology of setting blade rows using an optimisation procedure. Concurrent research in the intelligent control of individual vane settings, via agent technology, is showing the way forwards towards an intelligent geometry compressor (named here the IGC). The techniques described here will enable the optimisation of the vane settings to be determined, and then instantaneously achieved by the new control system.

Problems also arise in the design of fixed geometry compressors where design rules are sufficiently ill-defined that during the development of a high performance multistage compressor, several stagger settings will often be tested in order to achieve the desired objective [4, 6, 7, 8]. In high speed multistage compressors, it may well be necessary to vary the staggers of several stages to obtain these required results, which can become a multi-dimensional engineering optimisation problem. The conventional approach is empirically based on considerable engineering experience, and the methodology proposed in this paper, which includes numerical optimisation, presents an alternative approach.

As the first step of the project, a numerical approach modelling variable stator stagger was developed. The approach has been developed in a multistage environment, and it has two main elements, modelling the effect of stagger setting on individual stage performance and predicting overall performance and surge point. The stage performance modelling is based on an one-dimensional meanline method where correlations are used to introduce real flow effects. The method uses (experimental or predicted) stage characteristics at design (nominal) setting to generate characteristics at other settings. A stage-by-stage model is used to 'stack' the stages together with a dynamic surge prediction model such that compressibility throughout the compressor has been dealt with properly.

The approach was further incorporated with optimisation algorithms forming a numerical methodology for the optimisation of stator vane stagger settings which can define the blade setting of fixed and variable geometry compressors with reduced testing. In addition the method can investigate the potential benefits of introducing additional rows of variable setting vanes.

To date, little work has been reported on the application of numerical optimisation techniques to axial flow turbomachines, although Rao and Gupta [9, 10] did report about the optimum design of axial gas turbine stage by using interior penalty function methods, and Massardo and et al applied the similar technique to axial flow compressor design optimisation incorporating this with a meanline analysis [11] and a through flow code [12] respectively. In all the above work, the optimisation analyses were only performed at design point conditions for a single stage. An earlier paper by Garberoglio and et al [7] described work on numerical optimisation of stagger vane and bleed setting in multistage axial compressors, but it was based on considerable overall experimental data which is not usually available.

The numerical methodology presented in this paper for optimising stagger vane setting has been developed in a multistage environment, of which the optimisation analyses are performed for all expected operating conditions (flow rates and speeds). Notably, it has only used individual stage characteristics at nominal stagger vane setting (and speeds), which are typically made available from the manufacturer. A direct search method incorporating a Sequential Weight Increasing Factor Technique (SWIFT) algorithm was then used to optimise stagger setting. The objective function in this optimisation is penalised externally with a updated factor which helped to accelerate convergence. The methodology has been incorporated into a FORTRAN program and validated using the data from a 7-stage compressor. Several numerical examples are provided. Results have demonstrated that variable stagger is a powerful method to rematch stages and which can be used to improve desired overall performance. Parametric studies on SWIFT have also been conducted, where the algorithm showed numerical stability and fast convergence. Various numerical examples are provided. Further details and additional results are provided in Ref. [13] upon which this paper is based.

2. MODELLING

Varying upstream stator vane stagger changes the stage incident flow angle affecting the flow angle distribution through the compression system. To predict the effect of a change in stagger setting on overall compressor performance and surge conditions, it is therefore essential to begin by investigating the effect of a change in stator stagger setting on individual stages.

2.1 Modelling Stage Characteristics
The method as used in the present study was a meanline method proposed by Steinke [14] for numerical prediction of stage characteristics. The method can allow for the effect of a change in stagger settings and (or) in rotational speeds. It involves a stage stacking procedure and flow iteration techniques. Correlations are used to consider the real flow effects. The method can allow for a change in stagger setting together with a small variation in rotational speed. As datum data, stage characteristics for nominal conditions (nominal rotational speeds and stagger settings) are generally available (either from test or prediction), these data are used to predict the stage performance at other speeds and (or) stagger settings.

The first step involves establishing the nominal velocity diagram and obtaining the representative stage performance. From the input nominal conditions, rotational speed (N), stator stagger angle (β) and flow rate (W), using flow iteration and stage characteristics interpolation techniques, and Euler's equation, the nominal velocity diagrams at rotor inlet and outlet can be established, fig. 2 and 3; the non-dimensional performance parameters, (ϕ_d)

C556/025 © IMechE 1999

(flow coefficient), (ψ_d) (pressure rise coefficient) and (η_d) (adiabatic efficiency) can be obtained for this particular speed, stagger setting and flow rate. Additionally, parameters such as the relative velocities at rotor inlet (V_{2d}^1) and outlet (V_{3d}^1) can also be obtained, which are exploited to update the flow deviation as described in a later paragraph.

The calculation procedures for a change in stagger setting and a change in speed involve different mechanism and are considered separately as described below.

Change of Stagger Setting - to calculate the influence of stagger resetting by:

a) establishing the revised rotor inlet flow angle by noting that a change in stagger of $(\Delta\beta)$ implies a similar change in the inlet flow direction onto the downstream rotor. This change in flow direction can be used to establish a revised velocity diagram, as shown in fig. 2, from which an equivalent shift in flow coefficient $(\Delta\phi)$ can be calculated and the modified flow coefficient $(\phi = \phi_d + \Delta\phi)$ used to obtain the new stage performance. Assuming no change in deviation (i.e, $\Delta\delta = 0$), the same revised velocity diagram can then be used to calculate the rotor inlet and outlet relative velocities, V_2^1 and V_3^1 which together with relative velocities at nominal stagger setting (V_{2d}^1) and V_{3d}^1 are used in a correlation to update rotor outlet deviation.

b) it is assumed in (a), above, that resetting of the upstream stator only changes the absolute inlet flow angle of the downstream rotor ($\Delta\delta =0$), so the pressure rise coefficient (ψ) at the new setting can be found using the velocity diagram, Euler's equation, and interpolation of stage characteristics. The change in pressure rise coefficient can be calculated, $(\Delta\psi = \psi - \psi_d)$.

c) a similar assumption to that used by Camp and Horlock [6] was employed where it is assumed that stage efficiency still holds the values at nominal setting.

Change of Speed - to calculate the influence of speed variation by :

a) Assume there is a corresponding change in the axial component of velocity (V_a) with a change in rotational speed (N), so that rotor inlet flow angle holds unchanged, fig. 3, the axial velocity, $(V_a)_f (= \dfrac{V_a}{N} \cdot N_f)$, can then be obtained [13].

Use available information, such as velocity vectors, $(V_a)_f$ and $(U_{2t})_f$, the revised rotor inlet velocity diagram for a fractional speed can be established, fig. 3. If there is no change in rotor deviation (i.e. $\Delta\delta = 0$) subject to a change in rotational speed, the corresponding rotor outlet velocity diagram can also be established.

b) Regarding stage adiabatic efficiency, the same assumption as that used for stagger resetting is applied, namely the stage adiabatic efficiency holds its value of nominal (datum) speed operation.

The stage temperature rise, $(T_3 - T_2)_f$, is obtained by using flow iteration and Euler's equation, subsequently the change in pressure rise coefficient, $\Delta\psi \ (= \dfrac{C_{p.\eta d.}(T_3 - T_2)_f}{((U_{3t})_f \cdot \dfrac{N}{N_f})^2} - \psi_d)$, is

obtained [13].

c) When the compressor speed is changed, the stage flow coefficient (and subsequently flow range) can also be changed. To modify the assumption made in a) allowing for the resultant changes in stage flow coefficient or flow range, a correlation was introduced by Steinke to update the flow coefficient, and it is written as

$$\phi_f = \phi \left[1 + \left[\left(\frac{\phi}{\phi_d} \right)^{\frac{Nf}{N}} - 1 \right] \cdot \left(1 - \frac{N_f}{N} \right) \right] \qquad (1)$$

Where ϕ_f represents the updated flow coefficient, ϕ_d denotes the representative flow coefficient (at nominal speed, flow rate and stagger setting), ϕ is the flow coefficient at the datum or nominal speed line.

Rotor deviation angle The above discussions on stagger resetting and rotational speed variation both ignore the influences of a change in rotor incidence on its deviation. Considering this point, Steinke [14] introduced the following correlation to modify the deviation :

$$\Delta\delta = 10.\left(\frac{V_3^1}{V_2^1} - \frac{V_{3d}^1}{V_{2d}^1} \right) \qquad (2)$$

The above mentioned two correlations (equations 1 and 2) were originally designated for the single-stage compressors or that of limited number of stages, and they are not suitable for the present multistage compressors. Therefore modifications were conducted by the authors which are detailed in Ref. [13].

2.2 Predicting Surge and Overall Performance

Overall compressor performance at any stagger setting can be achieved by stacking the stage characteristics obtained using the methods described in the last section, however, the surge flow rate remains to be defined and although there are simple surge criteria [15] and incompressible surge models [16, 17], the authors feel that they are inaccurate in the multistage environment where compressibility and subsequent stage mismatching effects cannot be neglected [1] [13]. Therefore, the authors turned to other types of stability prediction method which are based on the stage-by-stage dynamic models where the compressibility is dealt with properly. Previous investigators obtained good surge predictions using this kind of criteria [18, 19, 20, 21, 22]. The technique involves: i) establishing the time-dependent mathematical models, ii) determining the flow rate at which instability (or surge) will occur.

A suitable computer code of this type (called 'Surge' in the later section) existed and was used in the present research. The background to such methods are described by Elder [18] and involve the following analysis:

i) a mathematical model is obtained by dividing the compression system into elemental volumes, as shown in fig. 4. Mass, momentum and energy conservation is applied to each element providing a set of dynamic equations as follows :

continuity: $\quad \int_{x1}^{x4} A(x).\frac{\partial \rho}{\partial t}.dx = W_1 - W_4$

momentum: $\quad \int_{x1}^{x4} \frac{\partial W}{\partial t}.\partial x = (W.V_a + P_s.A)_1 - (W.V_a + P_s.A)_4 + F_{net} \qquad (3)$

energy: $\quad \int_{x_1}^{x_4} A(x).\frac{\partial}{\partial t}\left[\rho.\left(C_v.T_s + V^2/2\right)\right].dx = C_p.\left[(T.W)_1 - (T.W)_4\right] + E_{net}$

F_{net} is the net axial blade force exerted on the flow and E_{net} is the effective work input. These two terms can be derived from stage characteristics based on quasi-steady assumptions [23]. In previous research, the stage characteristics were usually taken from interstage test measurements whereas the present study involves stagger settings where no experimental data are available and in this situation the experimental characteristics are used at the nominal conditions and at other settings the method described in the last section is used to obtain the corresponding performance. Various assumptions are made to solve the above equations and these are reviewed by Escuret [24].

ii) the surge criteria applied in the 'Surge' code uses a small perturbation analysis where the equations are linearised and represented as :

$$\frac{d \vec{X}_{dyn}}{dt} = C . \vec{X}_{dyn} \qquad (4)$$

$\vec{X}_{dyn}$ includes all the perturbations of the dynamic flow variables, while C represents a matrix containing geometrical constants and steady parameters. The surge point is then to be predicted once any of the eigenvalues of matrix C has a positive part. For fixed geometry study, the corresponding FORTRAN program were validated using the data from several multistage compressors, as shown in fig. 5.

3. OPTIMISATION

The objective of this process is to achieve the desirable (optimum) compressor performance by adjusting several stagger vanes. In practice, variations in stagger settings have upper and lower mechanical limits; achieving a certain peak performance usually involves specific application requirements. Thus, optimising stagger setting is a nonlinear constrained problem with multi-variables which can be generalised as follows :

$$\min f (\vec{x}) \quad \vec{x} : \text{n dimensional vector} \qquad (5)$$
$$\text{subject to : } g_i(\vec{x}) \geq 0. \quad I = 1, n_c$$

where n and n_c denote the number of optimisation variables and constraints, and $g_i(\vec{x})$ involves both nonlinear performance requirements and linear stagger variation bounds. Theoretically, these variation bounds can be as high as the mechanically permitted level. However, it is found from the present study that high level of bounds can bring the optimisation process beyond the model capacity, that is, flow iteration in variable stagger modelling becomes divergent. Therefore these stagger variation bounds are defined allowing for both mechanical limitations and the model capacity.

Solving this minimisation problem is to find an n dimensional parameter $(\vec{x})$ which minimises the general performance function and satisfies $g_i(\vec{x}) \geq 0$. In other words, the constraints, $g_i(\vec{x})$ construct a sub-space usually termed feasible region (or allowed region), and the (converged) optimum solution to the problem must stay within this region.

To solve the problem, the constraints must be dealt with properly. There are two different types of methods as commonly used, they are referred to as feasible methods and penalty function methods [25]. The former requests every trial search to stay within the feasible region, thus the constraint satisfaction checks are needed for every trial search. While the latter introduces the penalty function which involves both the objective function and constraints, it actually converts the problem into a non-constrained one and various optimisation methods for the non-constrained problems are therefore available. This kind of penalty function method has already been used in optimum design of axial flow compressor stages [11, 12] and also used in optimum design of axial gas turbine stages [9, 10] demonstrating the capability of dealing with constraints. Moreover, the penalty method consists of interior and exterior methods. The starting point for an interior method must lie inside the feasible region; while the exterior method can start outside the feasible region providing additional flexibility.

Many approaches have been exploited for solving the optimisation problem without constraints [25]. Those methods were classified into two categories by Schwefel [26]: direct (numerical) methods and indirect (analytical) methods. The direct methods search the solution in a stepwise manner (iteratively) by improving the values of the objective at each step. While the indirect methods attempt to reach the optimum in a single step or number of steps. The indirect method is based on analysing the special properties of objective function, such as the first and second-derivatives, but for practical problems such as that proposed here, these data are not yet available. Alternatively, some finite-difference approximate methods can be used, unavoidably, however, these will introduce some numerical errors, such as truncation errors and rounding errors [25, 27], and the solution become less reliable.

Mathematical software libraries, such as the NAG library, provide some generic algorithms suitable for general purpose applications, however, most routines need some derivative information for the objective function and nonlinear constraints, this information is not readily available for practical problems as discussed above and other routines also require the user to convert the constrained problem into a non-constrained situation before application. Because of these problems, the authors used the SWIFT algorithm which involves a combination of the exterior penalty function and direct (derivative-free) search methods and is described below.

SWIFT The method was described by Sheela and Ramamoorthy [28] as Sequential Weight Increasing Factor Technique. It exploits a unique technique to convert the constrained optimisation problem into non-constrained type incorporating the exterior penalty function (with a penalty factor of sequential increasing weight). Nelder and Mead's Simplex [29] algorithm was used to solve this non-constrained problem, which compares and then improves the function values at each vertex of a n+1 dimensional polyhedron for an n dimensional problem. The mathematical expression (5) can be now rewritten, as

$$\min P\left(\vec{x}, r_k\right) = \min f\left(\vec{x}\right) + r_k \sum_{i=1}^{n_c} \left| \min\left(g_i(\vec{x}), 0 \right) \right| \tag{6}$$

where $r_k = \max (1/d, r_{k-1})$, d is the mean values of distances from polyhedron centroid to each vertex, while r denotes the penalty factor, k and k-1 denote the k th and k-1 th iteration respectively. It is r_k that speeds up the convergence of algorithm.

The termination rule used is also that presented by Nelder and Mead (1965) where the variation of the penalty function values at the vertices of polyhedron is less than a prescribed value or that the polyhedron become small enough.

As pointed out by Gill [25] and others, poor scaling of variables can cause optimisation algorithm failure. The reason for this is that poor scaling can cause the objective function to be

C556/025 © IMechE 1999

ill-conditioned. To overcome the problem, non-dimensional stagger angles were introduced, that is

$$N_i = (\Delta\beta_i - \Delta\beta l_i)/(\Delta\beta u_i - \Delta\beta l_i) \qquad (7)$$

where $\Delta\beta$ denotes the variation of stagger angle, it is the independent variable; $\Delta\beta u$ and $\Delta\beta ul$ represents its upper and lower bound respectively. $|X_i|$ now takes a value between 0 and unity.

4. PROGRAM STRUCTURE AND IMPLEMENTATION

Based on the above analysis, a FORTRAN code was developed. It was written in structural blocks. Each block has its own specific function making it easy to modify (or replace). The program consists of the following routines : Main, SWIFT, Penalty, Objective, Surge, Stage and Constraint. Fig. 6 illustrates the program structure.

While implementing the program, the objective and constraints are needed to be specified by the user, which are detailed in Ref. [13] and briefed below.

Compressors often operate at a specific flow rate on each speed line, these flow rates are usually called operating conditions, (the corresponding performance on a compressor characteristic map constitutes the so-called operating or reference points and the optimisation was performed for these conditions. Usually the highest efficiency are desired at these conditions, although the balance between stable flow range and high performance depends on the application. The process used here can be used to optimise for the stall margin, performance, or mixed parameters. The following are common search requirements :

- optimise for maximum adiabatic efficiency with restrictions on pressure ratio and (or) surge flow
- optimise for maximum pressure ratio with restrictions on surge flow and (or) efficiency
- optimise for minimum surge flow with restrictions on pressure ratio and (or) efficiency

Results for each case are given in the next section.

5. RESULT

[1]**Case Study 1** - To obtain the maximum adiabatic efficiency at 100% design speed subject to the following constraints: $0 \leq \Delta\beta_4 \leq 2$, $0 \leq \Delta\beta_5$ and $\Delta\beta_6 \leq 3$, and $0 \leq \Delta\beta_7 \leq 5$. The optimisation process provides the results shown in table 1.

Table 1: Optimised Results (Case 1)

$\Delta\beta_4$	$\Delta\beta_5$	$\Delta\beta_6$	$\Delta\beta_7$	$\Delta\eta_{ref}$
1.010	1/495	2.984	5.012	2.68%

Comparisons of both stage and overall performance at design and optimum settings are presented in fig. 7 (a) and (b). The setting has improved the overall efficiency by almost 2.7% by improving the matching of the 7th stage. The consequent overall performances at 84% and 70% speed of the `newly staggered compressor' are also presented in fig. 7 (c) (together with optimised 100% speed curve), where some performance deterioration will be noted although there is a slightly improved surge-free flow range. It has been illustrated here that with a fixed

---

[1] Note: Case 1-3 uses the data from 7-stage LP compressor with hypothetical variable stator vanes; while case 4 uses the data from 12-stage HP compressor with variable IGV setting.

compressor geometry optimum stagger vane setting for design speed may inhibit part speed operation.

Case Study 2 -To obtain the minimum surge flow rate at 84\% design speed subject to the following constraints :

Optimisation 1: $0 \leq \Delta\beta_1 \leq 8$, $0 \leq \Delta\beta_2$ and $\Delta\beta_3 \leq 3$, and $0 \leq \Delta\beta_4 \leq 2$
Optimisation 2: the above and in addition $Pr \geq 2.63$

Table 2: Optimised Results (Case 2)

Setting	$\Delta\beta_1$	$\Delta\beta_2$	$\Delta\beta_3$	$\Delta\beta_4$	ΔW_{sg}
Opt. 1	7.60	1.05	1.50	0.40	-5.47
Opt. 2	6.00	1.05	1.50	0.40	-4.66

Comparisons of the results are presented in table 2. The overall performance subject to the two sets of conditions are presented in fig. 8 (a), where a significant improvement in surge-free flow range for both cases can be noted, but these associated with some reduction in pressure ratio and efficiency at operating flow. A smaller reduction in optimisation 2 will also be noted, because, in this case, the pressure ratio is constrained. The overall performance at 100% and 70% speed are presented in fig. 8 (b) (together with the 84% speed curve) for optimisation 1 setting, an improvement in surge-free flow range at 70% speed will also be noted with a small drop in pressure ratio. At design speed, the surge flow has been shifted to the left, but the compression system tends to be choking even at a much smaller flow rate. (Such phenomena have been used to obtain a wide operating flow range in industrial compressors, which are often driven by constant speed motors [15]).

Case 3 - To obtain the maximum adiabatic efficiency at 84% design speed subject to the following constraints : $0 \leq \Delta\beta_1 \leq 12$, $0 \leq \Delta\beta_2 \leq 8$, $0 \leq \Delta\beta_3 \leq 6$, and $0 \leq \Delta\beta_4 \leq 4$.

Fig. 9 (a) presents the optimised overall performance, where with the optimum stagger setting, an improvement in adiabatic efficiency and surge-free flow range is visible, but this is associated with a drop in pressure ratio. Fig. 9 (b) depicts the optimisation path. The minimum search is terminated at the 10th iteration; the efficiency, η, is ascending and the optimisation function (f and P) is declining. It is noted that all the trial search points have coincidentally fallen into the feasible region and the resultant penalty terms become zero, therefore the penalty and objective function, f and P, are the same. The penalty factor, r_k, is increased from the initial value of 2 to 37.31 at the 2nd iteration, and to 37.34 at the 3rd which was held till the end.

Parametric studies on the optimisation algorithm have also been conducted where it showed numerical stability and fast convergence [13] [2].

6. CONCLUSION

A method modelling variable stator vane setting is presented. This method was devised in a multistage environment and is capable of dealing with multistage compressors where the compressibility effect can not be neglected.

The combination of stage characteristics modelling, and overall performance and surge prediction formed a method for modelling variable geometry (stator vane) compressors and it was validated using the data from a multistage HP compressor.

C556/025 © IMechE 1999

To optimise the setting, a direct search method incorporating a Sequential Weight Increasing Factor Technique (SWIFT) algorithm was incorporated into the variable stagger modelling. The objective function in this optimisation is penalised externally with a updated factor which helped to accelerate convergence. The methodology has been incorporated into a FORTRAN program and its validations were conducted using the data from the multistage compressors.

From the applications of the optimisation code the following was observed: the instability and performance deterioration are caused by rear stages at high speed and by front stages at part speed operation. The optimum stagger setting at design speed operation does not provide good performance for part speed operation. With variable stator stagger, however, it is possible to obtain the optimum performance at both design and part speed operation. Also, varying the stagger of the front stages provides a powerful technique to rematch the stages in order to obtain a higher overall performance with wider surge-free flow range. The level of performance potential can be achieved by introducing extra variable geometry which is also well-demonstrated. Although such conclusions are already well established, the work presented offers a method of quantifying these benefits.

The code has showed numerical stability on parametric studies. It also has a fast convergence speed. It can provide a guide for the operation of compressors with variable stagger vanes and can also be used in compressor development programmes to determine the re-staggering of `fixed' vane settings.

Previous research shows that the actuators capable of shifting compressor characteristic can permit the design of robust controllers of effectiveness and simplicity [30, 31, 12]. In the present study, variable stagger vane techniques have quantitatively demonstrated the power in changing compressor characteristic and the developed models for variable stator vane setting therefore can be used for future consideration in designing active controllers.

REFERENCES

[1] I.J. Day and C. Freeman. *The unstable behaviour of low and high speed compressors.* ASME Paper 93-GT-26, 1993.

[2] J. Sun and R.L. Elder. *Numerical optimisation of stator vane setting in multistage axial flow compressors.* Proceedings of the Institution of Mechanical Engineers, Part A, 1998.

[3] M.T. Gresh. *Compressor Performance: Selection, Operation and Testing of Axial and Centrifugal Compressors.* Butterworth-Heinemann, 1991.

[4] N.A. Cumpsty. *Compressor Aerodynamics.* Harlow : Longman Scientific and Technical, 1989.

[5] D.E. Muir et al. *Health monitoring of variable geometry gas turbines for the Canadian Navy.* ASME Journal of Engineering for Gas Turbines and Power, 111, April 1989.

[6] T.R. Camp and J.H. Horlock. *The analytical model of axial compressor off-design performance.* ASME Paper 93-GT-96, 1993.

[7] J.E. Garberglio, J.O. Song, and W.L. Boudreaux. *Optimisation of compressor vane and bleed settings.* ASME Paper 82-GT-81, 1982.

[8] A. Sehra, J. Bettner, and A. Cohn. *Design of a high-performance axial compressor for utility gas turbine.* ASME Journal of Turbomachinery, 1994, April 1992.

[9] S.S. Rao and R.S. Gupta. *Optimum design of axial flow gas turbine stage, part i: Formulation and analysis of optimisation problem.* ASME Journal of Engineering for Power, 102:782-789, 1980.

[10] S.S. Rao and R.S. Gupta. *Optimum design of axial flow gas turbine stage, part ii: Solution of the optimisation problem and numerical results.* ASME Journal of Engineering for Power, 102:790-797, 1980.

[11] A. Massardo, A. Satta, and M. Marini. *Axial flow compressor design optimisation. Part i: Pitchline analysis and multivariable objective function influence.* ASME 89-GT-201, 1989.

[12] A. Massardo and A. Satta. *Axial flow compressor design optimisation. Part ii: Through-flow analysis.* ASME 89-GT-202, 1989.

[13] J. Sun. *Modelling Variable Stator Vane Setting in Multistage Axial Flow Compressors.* PhD Thesis, SME Cranfield University, Bedfordshire, England, 1998.

[14] R.J. Steinke. *Stgstk: A computer code for predicting multistage axial-flow compressor performance by a meanline stage-stacking method.* NASA TP 82-2020, 1982.

[15] A.B. McKenzie. *Axial Flow Fans and Compressors: Aerodynamic Design and Performance.* Cranfield Series on Turbomachinery Technology. Ashgate Publishing Limited, 1997.

[16] E.M. Greitzer. *Surge and rotating stall in axial flow compressors, part i: The- oretical compression system model.* ASME Journal of Engineering for power, 98, 1976a.

[17] F.K. Moore and E.M. Greitzer. *A theory of post-stall transients in axial compression systems: part i and ii.* ASME Journal of Engineering for Gas Turbines and Power, 108, 1986.

[18] R. Elder. *Mathematical Modelling of Axial-flow Compressor.* PhD thesis, University of Leicester, England, 1972.

[19] W.A. Tesch, R.H. Moszee and W.G. Steenken. *Linearized blade row compression component model, stability and frequency response analysis of a j85-13 compressor.* NASA CR 135162, 1976.

[20] M.E. Gill. *Surge Prediction in Multistage Axial and Centrifugal Compressors.* PhD thesis, Cranfield Institute of Technology, England, 1986.

[21] M.W. Davis and W.F. O'Brien. *A stage-by-stage post-stall compression system modelling technique.* AIAA 87-2088, 1987.

[22] J.F. Escuret and R.L. Elder. *Active control of surge in multistage axial-flow compressors.* ASME 93-GT-39, 1993.

[23] A.G. Corbett and R.L. Elder. *Stability of an axial flow compressor with steady inlet conditions.* ASME Journal of Engineering for Power, 16, 1974.

[24] J.F. Escuret. *The Prediction and Active Control of Surge in Multi-stage Axial-flow Compressors.* PhD thesis, SME Cranfield University, Bedfordshire, England, 1993.

[25] Philip E. Gill, Walter Murray and Margaret H. Wright. *Practical Optimisation.* Academic Press, 1981.

[26] Hans-Paul Schwefel. *Numerical Optimisation of Computer Models.* John Wiley & Sons, 1981.

[27] R. Fletcher. *Practical Methods of Optimisation.* A Wiley-Interscience Publication, second edition, 1987.

[28] B.V. Sheela and P. Ramamoorthy. *Swift - a new constraint optimization technique.* Computer Methods in Applied Mechanics and Engineering, 6:309-318, 1975.

[29] J.A. Nelder and R. Mead. *A simplex method for function minimization.* Comp. J., 7:308-313, 1965.

[30] R.L. Behnken and R.M. Murray. *Combined air injection control of rotating stall and bleed valve control surge.* In Proc. American Control Conference, 1997.

[31] S. Yeung and R.M. Murray. *Nonlinear control of rotating stall using axisymmetric bleed with continuous air injection on a low-speed, single stage, axial compressor.* In Proc. Joint Propulsion Conference, Seattle, WA, July 1997.

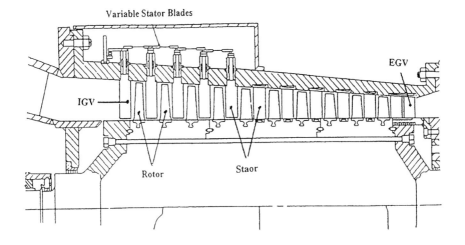

Figure 1: Axial compressor with variable stators (from Gresh, 1991)

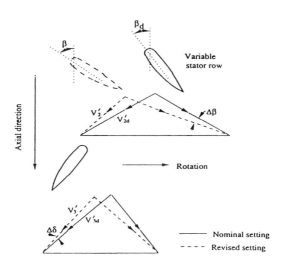

Figure 2: Rotor Velocity Diagram with Variable Upstream Stator Vanes

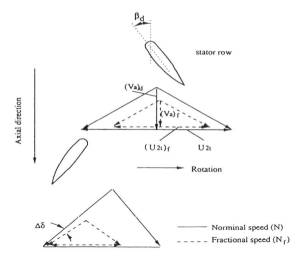

Figure 3: Rotor Velocity Diagram with Variable Rotational Speed

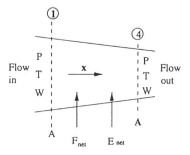

Figure 4: Elemental Volume

C556/025 © IMechE 1999

7 - Stage LP Compressor

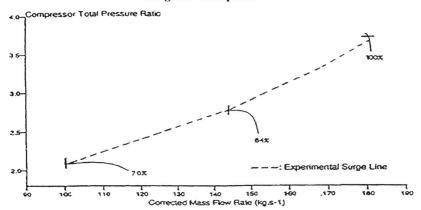

7 - Stage HP Compressor with Variable Geometry

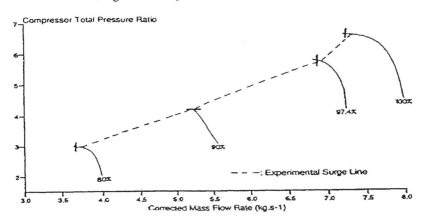

Figure 5: Validations of "Surge" Code (from Escuret, 1993)

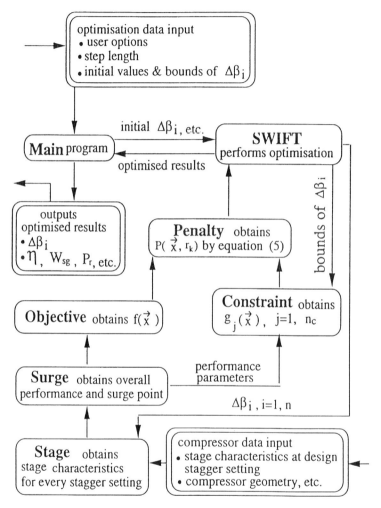

Figure 6: Program structure

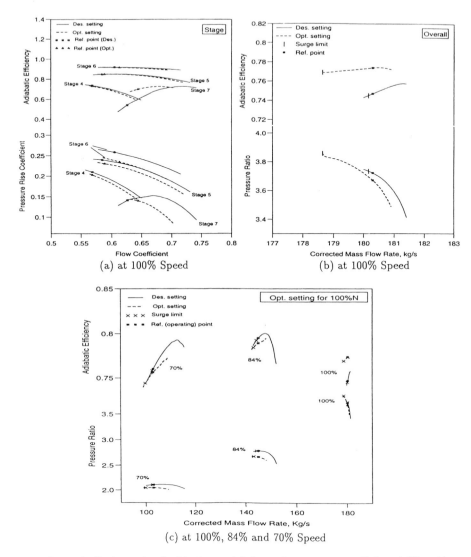

Figure 7: Optimisation for Maximum Adiabatic Efficiency at 100% Speed (Case 1)

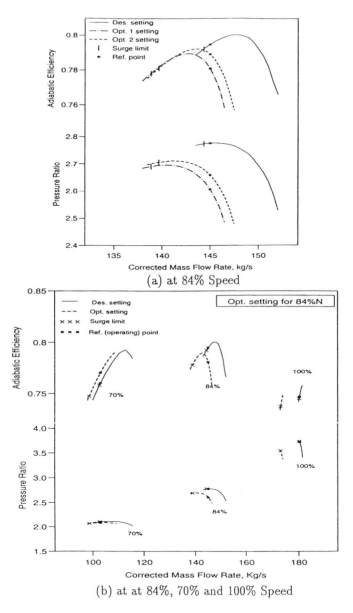

(a) at 84% Speed

(b) at at 84%, 70% and 100% Speed

Figure 8: Optimisation for Minimum Surge Flow Rate at 84% Speed (Case 2)

C556/025 © IMechE 1999

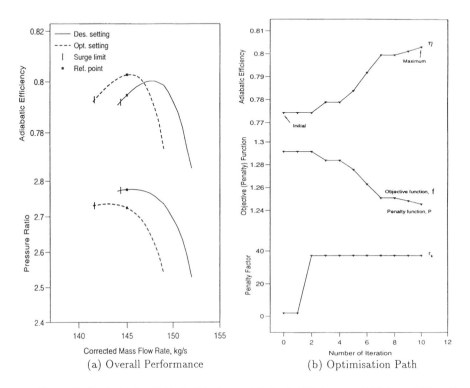

(a) Overall Performance   (b) Optimisation Path

Figure 9: Optimisation Path for Maximum Adiabatic Efficiency at 84% Speed (Case 3)

C556/012/99

Demonstration of a subsea multiphase flow pumping system

J M KUJAWSKI
Westinghouse, Electro-Mechanical Division, Pennsylvania, USA
E F CAETANO
R&D Center, Petrobras, Rio De Janerio, Brazil

Cc. WE Harden
W Vance
NRA Munly
D Broadhurst
Lm Bilby

SYNOPSIS

The Westinghouse Electro-Mechanical Division (WEMD), USA and PETROBRAS, Brazil, have been working on the design and manufacture of a unique combination of existing technologies and novel ideas to boost multiphase crude from subsea wells. As part of the PROCAP-2000 program, the first system, titled the SBMS-500 by PETROBRAS, and the WELLAMPS ™ multiphase pumping system by Westinghouse, will provide up to 500 m³/hr total flow rate and 60 bar of pressure increase, for all GVF (gas void fraction) conditions up to 95%, volumetric. This paper presents the technological and economic basis for the concept, and the Technology Implementation Project (TIP) qualification program.

1. INTRODUCTION

The Westinghouse Electro-Mechanical Division (WEMD), USA and PETROBRAS, Brazil, as part of the PROCAP 2000 program, have teamed to design and manufacture a submerged multiphase pumping system that will boost multiphase crude from subsea wells in a reliable and economical fashion. The system design is complete, and final manufacture is in progress for delivery in 2Q99 to the land based multiphase flow test facility at the Atalaia Production Complex, near the city of Aracaju, in the northeast region of Brazil. KVAERNER is a key team member for the deployment manifold, subsea control module, and other components including a significant portion of the instrumentation. This paper presents the technological and economic basis for the concept, pictures of the manufactured components, and the details of this Technology Implementation Project (TIP) planned qualification program. Once qualified, this system will be the basis for a family of designs that will be capable of producing high flow rate crude from deep water wells, over long distances, thus providing oil companies the opportunity to enhance existing wells, as well as extend the life of low performing wells. The system offers oil companies several new applications and augments their operational flexibility, either in hostile or milder environments, resulting in significant Capital Expenditures (CAPEX) and Operational Expenditures (OPEX) savings.

The focus of this paper is the subsea motor-pump set module, and the top-side support systems, which are being supplied by Westinghouse (Figures 1 and 2). Within this paper, these systems will be referred to as the SMPS (Subsea Multiphase-Pumping Subsystem).

By combining a reliable <u>canned</u> motor design and a Leistritz twin-screw pump, with innovative, patented pressurizing equipment, the SMPS packages existing top-side products into a reliable subsea system. The SMPS design, manufacture and factory component testing are complete, and qualification testing of the completely assembled SMPS is scheduled to begin in 3Q99 at the Atalaia test Site. After completion of 1000 hours of engineering testing, the SMPS will be inspected and then deployed in Marlim Field of the Campos Basin.

The following paragraphs from Reference 1 provide an excellent description of successful trials to demonstrate screw pump multiphase pumping technology.

> Multiphase fluid has been defined many different ways. In oil and gas production, the term two phase flow is used to describe a mixture of hydrocarbon liquid and hydrocarbon gas. If water is added as a separate immiscible phase, the system is described as three phase or multiphase flow.

> Oil wells produce a mixture of oil, water, and gas, and occasionally sand, natural gas hydrates and waxes. The transfer of this mixture via a flow line to a central processing facility is called a multiphase production system. Since the inception of oil production, multiphase fluid has been transferred over short distances using reservoir energy. There was no device available to directly pressure boost the multiphase fluid when reservoir pressure was insufficient. The only feasible approach was to separate the phases and independently use a pump for the liquids and a compressor for the gases. The machine to perform that duty, the multiphase pump, has finally become a reality.

> In late 1992, Chevron joined the Texaco Joint Industry Program (JIP) in which several types of multiphase pumps were tested in their Humble flow loop in Houston. This facility allowed hydrocarbon mixtures of oil, gas, and water to be measured and delivered to the pump suction for pressure boosting in any proportion desired. Test results were issued to the members.

> Participating in this program allowed efforts to be focused in several ways. Based on the JIP results, Chevron decided to pursue twin screw positive displacement pumps instead of the other contenders like the hydrodynamic helical axial pump (born out of the Poseidon Joint Venture between Total, Statoil, and IFP - the French

Petroleum Institute). Chevron also decided in late 1993 to procure two twin screw multiphase pumps for two field trial programs (one low pressure boost test, and one high pressure boost application). By running these pumps under actual field conditions, Chevron would gain the knowledge necessary to make an informed decision about the future of this technology and its application to Chevron assets. Additionally, operational experience would be gained by Chevron personnel.

The following paragraphs, extracted from Reference 2, provide a brief description of the Petrobras environment and evolution of their planning to produce oil from these

At present in Brazil, the reserves accumulated in fields located in waters of 400 to 1,000 meters (classified as deep) and those in depths over 1,000 meters (classified as ultra-deep) account for approximately 64% of Brazil's reserves. In addition, the Company exploratory play-analysis indicates that over 65% of the potential oil discoveries (prospective discoveries) will happen in deep and ultra-deep waters. These facts and figure values demonstrate how important the deep and ultra-deep waters scenario is to Brazil.

In 1986, in the search for the technological capability necessary to exploit the deep water fields already discovered, PETROBRAS established a technological strategic program called PROCAP (PETROBRAS' Technological Development Program on Deep Water Production Systems). This six-year, US $70 million program - which ended in 1991 - combining technological actions involving PETROBRAS, other petroleum companies, service companies, R&D institutions and universities, in the country and abroad, has allowed, under the Company demanded time and cost frame, full technological capability to exploit oil fields at water depths (WD) up to 1,000 meters. This program success was crowned with Marlim-4 well completion at 1,027 m WD in early 1994.

In 1992, having many discovered fields stretching beyond the 1,000 m, even crossing the 2,000 m water depth boundary, and considering the excellent results obtained with the PROCAP, PETROBRAS decided to create a new program named PROCAP-2000. (Technological Innovation Program on Deep Water Exploitation Systems). This Program, far more daring than the previous one, as it pursues the development of many more innovative technologies, and is intended to lead the Company to the full technological capability of exploiting oil fields at WD up to the 2,000 meters boundary, making use of technologies that introduce strong changes on the current methods for deep water production.

It is from these considerations, as well as other information within references 1 and 2 and other demonstrated operation of equipment, that a Cooperative Agreement was created to develop and demonstrate the use of a full size multiphase pumping system.

The remainder of this paper will describe the design basis, factory test results, future land based and subsea test plans, and expected economic benefits of the SMPS technologies.

2. DESIGN BASIS

2.1 Philosophy

The design philosophy for the SMPS was:

• Use demonstrated technologies, in a uniquely packaged configuration.
• Use a single fluid to serve all of the system lubrication functions, including heat transfer.
• Provide a positively pressurized clean lube oil system, to assure clean oil flow through the pump seals, and maintain a clean system to assure long life for the system bearings and gears.

Thus, the intent was to reduce complexity, and enhance reliability using features and systems with demonstrated reliability and longer life.

Once this design concept feasibility was established, PETROBRAS and WEMD agreed to: complete the design and manufacture of a full scale system, test the components individually, test the system at a land based test site, and then deploy it subsea at a depth of 660m (design depth of 1000 m) to pressure boost a flowing well.

A complete package of specifications for the system was generated collectively while performing the design effort. With the specifications defined, a family of designs, under the trademark WELLAMPS™, capable of covering a wide range of flow rates, pressure boost, gas void fraction, and fluid viscosities, was conceived.

2.2 Design and Manufacturing

2.2.1 Subsea Set

The **Pump** selected for this application, is a self-priming positive displacement rotary screw pump. Reasons include the testing knowledge cited in reference 1, and this pump's unique ability to handle a wide range of gas void fraction and viscosity applications. Figure 3 is a photo of the pump while on factory test. Designed for this first system production application, (500 m3/hr (2200 gpm) flow rate, and up to 60 bar (870 psig) pressure increase), it is ideally suited for multiphase pumping of fluids from low to high viscosity, and gas fractions up to 100%, which can be aggressively corrosive and abrasive. The rotary pump design consists of a fixed casing containing two rotary screws, bearings, seals, and timing gears. Operating history has shown that long life can be anticipated when the seals, bearings, and gears are properly lubricated, and the system is designed to do this.

The **Motor** for the prototype system is a canned induction motor, but unique in that the design incorporates a novel insulation system that allows the use of medium level voltage up to 6.9kV at the motor terminals. This eliminates the need for a subsea transformer, while at the same time providing the ability to use a reduced size conductor on the power umbilical, and allowing the use of standard umbilical materials.

The canned motor design was selected because it provides all of the known reliability benefits of separating the motor windings from the process fluids in critical applications. With the motor hermetically sealed, such designs have reliably performed for greater than 20 years without maintenance. Figure 5 shows an illustration of a typical canned motor. (Reference 3 provides a more detailed description of canned motors.) Briefly, the rotor and stator electrical components are protected by a hermetically sealed metal can…thus the term, "canned motor". This architecture of the canned motor makes it intrinsically more reliable than a conventional fluid (either oil or water or gas) pressurized motor type when subsea deployed because the can is impermeable to the diffusion of gases or liquids. Literature reports that only a few parts per million (ppm) quantities of water invading a fluid pressurized motor is sufficient to provoke a motor electrical failure. Typical of most induction motors, radial bearings are positioned on both ends of the rotor core. A thrust bearing is positioned to take system thrust loads, and end caps and shells are provided to contain the various motor operating pressures and temperatures. Various methods are used to cool the units, including internal auxiliary impellers that circulate the fluid, through heat exchangers. The electrical connection is made through a hermetically sealed terminal gland.

The SMPS motor is a 3 phase, 4 pole unit capable of operating with a variable frequency input from 20 to 60 Hz. The motor is rated at 1268kW (1700 hp). The unit is designed to operate with a high current density, which reduces required winding volume. The motor is slender in profile to reduce fluid drag losses.

The main power connectors are a dry mate-able rated at 11kV and 200 Amps, and a water mate-able one rated at 8kV and 200 Amps. These connectors are used with a Subsea Optical Power Umbilical, which will be deployed in a free catenary shape.

The other ancillary subsea equipment includes the **Heat Exchanger** and the **Pressurizer System**, which are both part of the novel, patented lubrication-cooling-sealing system. The heat exchanger supplies supplemental cooling for the subsea motor pump set. In future systems, the residual heat may be added to the produced fluid,reducing potential for wax formation. The pressurizer provides positive and bounded pressure above wellhead pressure and pressure fluctuations. Coupled to the topside Oil Supply Unit (OSU), the system provides and maintains a clean fluid environment for all of the subsea component bearings and gears. It also pressurizes and flushes the pump seals, thus providing long life and high system reliability.

Complementary Sub-systems are designed and supplied to house, deploy, and connect the subsea system to the topside equipment. The design employs techniques specifically developed for deployments in deep waters, such as guidelineless and vertical connection operations. Some operations are being designed to be performed by simple and low operational cost surface vessels. Some of these items are illustrated in Figure 4.

2.2.2 Topside Equipment

Figure 2. shows the topside equipment, consisting of the Master Control Station (MCS), the Variable Frequency Drive (VFD) and its Enclosure, and the Oil Supply Unit (OSU).

The **Master Control Station** is linked to all of the SBMS-500 system components and instrumentation. Its purpose is to:

- Control routine and continuous activities with the VFD, OSU, the platform interfaces, and the Subsea pumping unit.
- Protect the subsea set from external problems.
- Limit consequences of any subsea set unusual operation.
- Provide performance information.

The system consists of a programmable logic controller (PLC) with an industrial personal computer (PC) to support the man-machine interface, permitting operator intervention and operation of the system. The MCS provides logic control of the Oil Supply Unit (OSU). The MCS communicates with and controls the Variable Frequency Drive and the Subsea Control Module (SCM). This SCM multiplexes the subsea instrumentation output, and transmits these signals to the Master Control Station, through the Subsea Optical and Power Umbilical. This implementation will be a pioneering application of a water mate-able optical connector. In addition, the Subsea Control Module provides the important function of controlling the lube oil inventory. Finally, the Master Control Station is slaved to the platform emergency shutdown system (ESD).

The **Variable Frequency Drive (VFD)** is a 7.2kV (rated) device that provides a power output of 1650kW. The VFD has the unique ability to supply rated power at very low harmonic distortion, meeting IEEE standard 519, and also providing low harmonic distortion waveforms to the motor and subsea power umbilical. This feature is important for this and all subsea electrical power systems, because it allows the components to operate at full design rating. The suppliers Perfect Harmony™ Drive Technology has recently demonstrated capability for much longer distances than the 2.5km necessary for this unit. This provides the possibility of remote placement of the subsea system from the power supply equipment, thus opening many opportunities for economic evaluations of multiphase pumping.

The VFD will be enclosed in a Det Norske Veritas (DNV) certified enclosure, classified to be located in an IEC 79, Group II a, Zone II, Temperature T-3, hazardous environment. It will be supplied as a packaged system, including all of the required electrical interconnections for installation onto the floating platform in the Marlim deepwater field.

The **Oil Supply Unit (OSU)** is designed to automatically supply pressurized lubrication oil on demand to the subsea set. The mechanical equipment of the OSU is self contained on a skid module that will be mounted on the main deck of the platform. Control functions will be provided from the Master Control Station. This unit will also be Det Norske Veritas certified.

The OSU is designed to be installed, exposed to the weather, in an IEC-79, Group IIA, Zone 2, Temperature T3, hazardous environment. The primary components of the OSU include an oil storage tank of 1000 gallons capacity, two electrically driven gear-type pumps, each capable of pumping high pressure oil at the required flow rate, an accumulator, and a dual element, hydraulic filter. Both manually operated and solenoid actuated valves are provided at various locations to allow venting of sections of the OSU piping and servicing of the fluid system components. During operation the OSU supplies oil into an umbilical which leads from the host platform to the subsea equipment. The OSU capacity is designed to operate continuously for long periods of time without replenishment.

3. TEST RESULTS

3.1 Factory Tests

3.1.1 Pump

Figure 3 shows the pump installed in the test stand for the factory testing.

The pump was tested using water as the test medium, per API STANDARD 676. A lube oil system was used to pressurize the seal system, and lubricate the bearings and gears. Pump head, flow, and power are supplied as Figures 3a, 3b, and 3c.

Post-test inspection of the unit verified the acceptability of the unit for subsequent testing with the system without modification.

3.1.2 Motor

Figure 6 shows the completed motor winding.

The winding final electrical tests were:
- 10 minute insulation resistance at 1000 Vdc.
- Terminal to terminal resistance
- DC high potential to ground at 24.7kVdc (momentary)
- AC high potential to ground at 14.5kVac (momentary)
- 1 minute insulation resistance test at 1000 Vdc.

A total of nine sets of electrical tests were successfully conducted during the manufacture of the motor, and a series of coil manufacturer proof tests were conducted prior to motor winding. These tests included a series of dc and ac production tests to 33.3 kVdc and 19.6 kVac. Destruct tests were also conducted.

The motor will be tested at no load and at locked rotor conditions using oil, to IEEE 252. In addition, the proper pressure, vibration, and leak testing will be performed. Figure 7 defines the calculated motor performance at 60 and 20 hz operation.

3.1.3 VFD

Testing of the SBMS-500 VFD took place in July of 1998 over a period of three days. Testing included: confirmation of drive rated current and rated voltage output capability, full load heat run (approximately 24 hours at full load), efficiency measurements, and waveform distortion measurements. All measured values met or exceeded specification requirements. Table 1 summarizes some of the important test results:

Table 1. VFD Factory Test Results

Frequency (Hz)	60	60.9	20
Current (Amps)	200	200	152
Volts (kV)	4	6.9	2.3
THD* (%), voltage	-	2.6	3.5
THD* (%), current	-	2.1	3.4
Efficiency (%)	98.2	-	-

* Total Harmonic Distortion

3.2 Land Based Test Plan

The Atalaia Production Complex is designed to test major components for deep water production systems. It is being modified to accept the SMPS system for complete land based engineering evaluative system testing prior to preparations for subsea well deployment.

The Atalaia test objectives are:
- Demonstrate the pump/motor performance
- Demonstrate the performance of the pump/motor sub-systems (seal inventory, heat exchanger, and stand-by pressure controls)
- Demonstrate appropriate interaction between components
- Test transient conditions
- Perform 1000 hours of operation and perform successful disassembly and inspection.

Multiphase dead oil with entrained gas at controlled levels will be used as the process fluid.

Table 2 (from reference 2) summarizes this qualification test program.

                                                         C556/012 © IMechE 1999

Table 2. - SBMS-500 Qualification Testing Program at PETROBRAS Atalaia Test Site.

	Preliminary Tests
A	Pump and Motor Alignment
T	Direction of Rotation
A	Electrical and Control Connections
L	Piping Hook-up and Cleanliness Check
A	Monitoring System Check
I	Load Slip Test
A	**Acceptance Tests**
	Water Performance Test
T	Live oil (recombined) Operational Test
E	Motor-Pump and Drive Starting Tests
S	Motor-Pump and Drive Acceleration Tests
T	Motor-Pump and Drive Deceleration Tests
	Motor-Pump and Drive Electrical and Hydraulic Performance Characteristics
S	Motor-Pump and Drive Electrical Tests
I	Instrumentation Operability and Vibration Monitoring
T	Guarantee Point Test
E	Extended Operation Test (1,000 h @ 170 to 500 m³/h)

Instrumentation is provided to monitor the pump, motor, heat exchanger, lube oil, variable frequency drive, and computer systems. Additional test instrumentation and monitoring equipment will be used during this land based testing to confirm auxiliary system acceptability and performance.

3.3 Marlim Field Test

The host well, 7-MRL-7-RJS, is located around 4 km southeast of the designated host platform. It was drilled in May, 1991, and completed as an oil producer in August, 1991. The most appropriate location for the system was selected by examining well characteristics including total flow rate, GVF at the suction, and pressure boost. The expected operating conditions will provide a good test of the system performance. Installation of the SBMS-500 could provide as much as 60% peak increase in the oil production. The tests shown on Table 3 are planned for the system:

Table 3. Marlim host well 7-MRL-7-RJS Qualification Test Program

M	**Prove-out Test**
A	After Deployment and Tie-backs
R	After One Month Operation
L	**Endurance**
I	24 mos. operation & reporting every 6 mos.
M	**Monitoring Characteristics**
S	Pump Performance (capacity, pressure boost and efficiency)
I	VFD Performance (voltage, current wave forms and delivered power)
T	Motor-Pump Set Vibration Monitoring
E	Instrumentation Operability and Performance

4. PROJECTED ECONOMIC BENEFITS

The major positive impacts and benefits of the usage of this novel technology of subsea multiphase flow pumping systems, are:

- Production anticipation by increasing well flow rates.
- Potential higher reservoir recovery factor.
- CAPEX reduction by requiring fewer wells to achieve a given production volume.
- CAPEX reduction by either consolidating surface facilities, or transporting the production over longer distances to existing infrastructures (allowance of long tie-backs).
- OPEX reduction by the simplification in the required production infrastructure and by achievement of higher availability because of system reliability.
- Potential reduction in the flowline network by possible commingling of the production of separate areas.
- Ability to produce profitably from marginal accumulations using long distance tie-back to existing infrastructures that offer spare power and excess process capacity.
- Allows even earlier deployment of Anticipated Systems for discovered field appraisal.
- Allows production where gas flaring is not allowed or is uneconomical to produce separately from the oil.
- Permits a better crude blend (commingling production of diverse areas) that results in easier field primary processing and higher price value for the blend.
- Maintenance of a constant and predictable well flow rate to the processing facilities (leverage of the natural well fluctuations).

Illustrating these impacts and benefits, the pioneer deployment of the System in the Brazilian Marlim deep water field is expected to provide the following specific results:

- A 60 % higher oil flow rate in the host well , resulting in significant oil production anticipation.
- Higher percentages of accumulated fluid recovery factor during the well life, i.e.,
  - Oil increased by 7 %
  - Gas increased by 10 %
  - Water increased by 14 %

- Improved Positive Net Present Values (NPV) even though limitations exist in the P-20 host unit in processing larger fluid quantities.

These results are expected to be achieved while the systems operation is governed by the usual petroleum production practices such as: no free gas generation is allowed at the well perforation region (i.e., sets a limit to the pump suction pressure value); no sand production is allowed (limits the well pressure draw-down); and, the reservoir fluids have to be compulsory replenished (reservoir pressure maintenance; subsidence effects not allowed).

In summary, within a fairly wide range of conditions, multiphase pumps can effectively pressure boost fluids for transport to larger processing facilities eliminating or reducing the need for separation, single phase pumps, and gas compression facilities within outlying productions centers. Multiphase pumps can enhance production from wells normally limited by first stage separator pressure and permit the economical development of satellite fields at greater distances from a central processing facility. In the near future, subsea multiphase pumping in deep water will be commonplace and reliable, bringing a new economic dimension to the development of satellite oil fields.

5. JOINT INDUSTRY PROJECT - (JIP)

Based on two main principles, aggregating technical competence and applying new technology as soon as possible, the theoretical and experimental actions imbedded in the non-exclusive Technological Cooperation Agreement was offered to the industry under a Joint Industry Project (JIP) format. The response has been very positive and now includes the following companies: AMOCO (USA), CHEVRON (USA), MARATHON (USA) and ORYX (USA). In addition to contributing to the program activities in regular program briefings and updates, the JIP members receive design and experimental result reports, and are welcome to witness all the phases of the Technology Implementation Project , which includes Atalaia land based testing, and the Marlim field testing.

6. CONCLUSION

The novel concept of a subsea multiphase flow pumping system is currently being developed. It will offer oil companies a very versatile tool that will positively impact the daily operation of the industry, especially in hostile environments (e.g. deep water) and/or when facing offshore marginal accumulations.

The first system is a full sized production system, applying sufficient power (1.2 MW) to a well to be able to produce 75,000 bbl per day. At this time the design efforts are complete and the manufacturing is on schedule for shipment for land testing in 3Q99.

Designs are being developed to serve the projected deep-water needs of the future. Figure 8 shows the planned performance range. Such systems will be able to handle all ranges of gas void fraction and viscosity pumping, with capability of water depths up to 3,000m.

ACKNOWLEDGEMENTS

The authors are extremely grateful for the outstanding contributions and efforts of the design teams at WEMD and Petrobras.

In addition, we acknowledge the efforts of design engineers at Leistritz, Kvaerner, Robicon, Pirelli, Tronic, Hebeler, EG&G Belfab, Sonic Barrier Sound Products, and Ocean Design Inc., for key component design and interfaces with the design teams.

Finally, we are grateful to the JIP members who have contributed significantly to the design through their shared industry experiences.

LIST OF REFERENCES

1. "MULTIPHASE PUMP FIELD TRIALS DEMONSTRATE PRACTICAL APPLICATIONS FOR THE TECHNOLOGY", D.F.Dal Porto, SPE, and L.A.Larson, Chevron Petroleum Technology Company, Oct, 1996, SPE 36590

2. "SBMS-500: A PETROBRAS and WESTINGHOUSE Cooperation on a Subsea Multiphase Flow Pumping System", E.F.Caetano, et.al., May, 1997,OTC 8454

3. "PUMP DESIGN - NEW BUT OLD", J.M.Kujawski, WORLD PUMPS, JULY 1993.

Westinghouse Motor/Pump Set Subsea Module

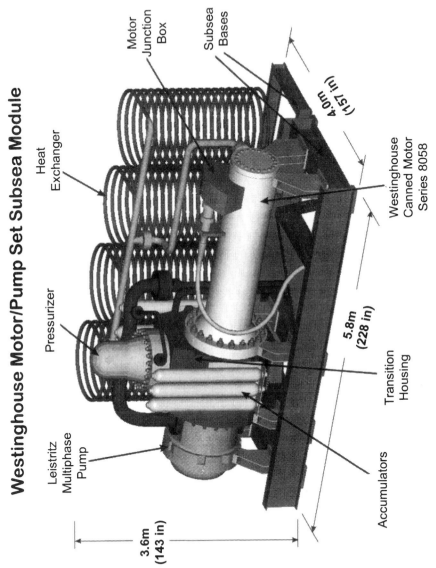

Figure 1. Motor/Pump Set Subsea Module

67

SMPS-500 Topside Components

VFD (In Container)   Master Control Station   Oil Supply Unit

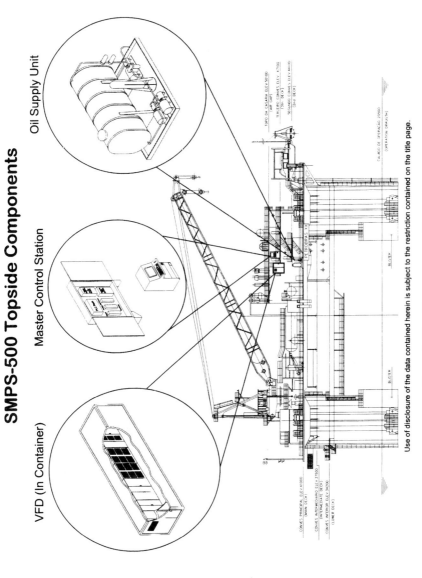

Use of disclosure of the data contained herein is subject to the restriction contained on the title page.

Figure 2. Topside components illustration.

Figure 3. Leistritz screw pump during factory testing.

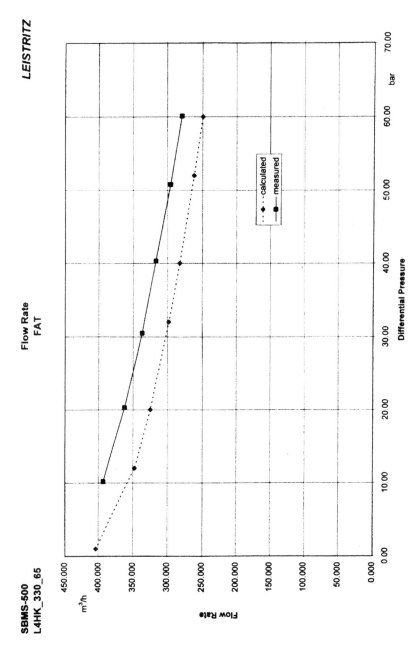

Figure 3a. Pump factory test results.

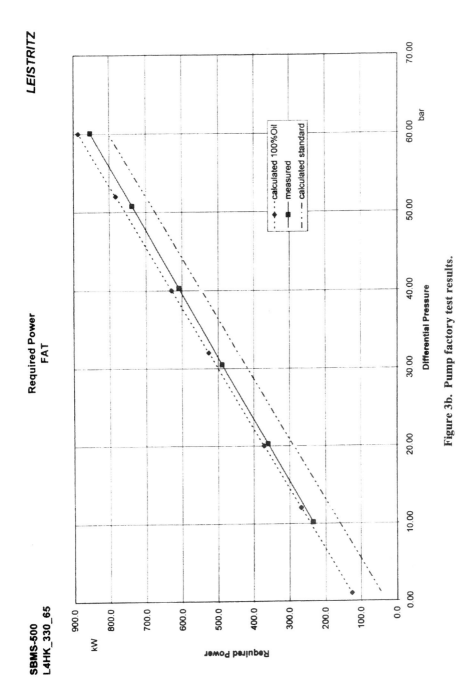

Figure 3b. Pump factory test results.

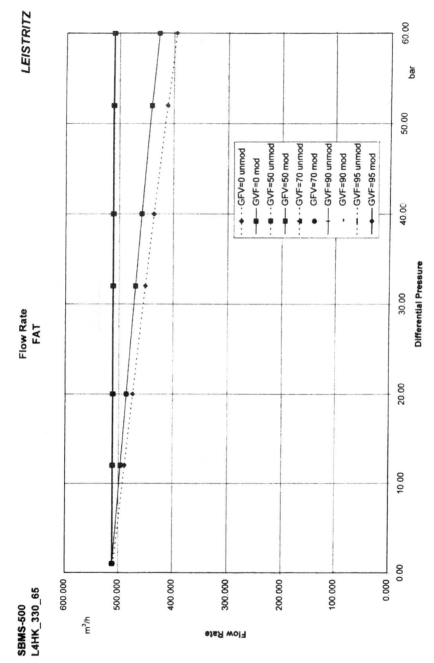

Figure 3c. Pump factory test results.

Figure 4. Kvaerner's mini-manifold deployment module.

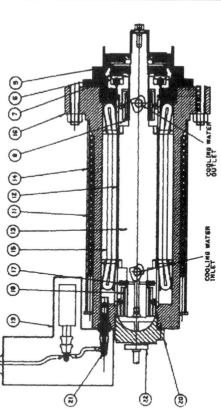

BILL OF MATERIALS	
ITEM	DESCRIPTION
1	NOT PICTURED
2	NOT PICTURED
3	NOT PICTURED
4	NOT PICTURED
5	SECONDARY IMPELLER
6	THRUST RUNNER
7	THRUST BEARING
8	RADIAL BEARING
9	PURGE LINE
10	FLANGE BOLTS
11	COOLING COIL
12	STATOR CAN
13	ROTOR ASSEMBLY
14	COOLING WATER JACKET
15	STATOR ASSEMBLY
16	STATOR SHELL & FLANGE
17	AUXILIARY IMPELLER
18	RADIAL BEARING
19	TERMINAL BOX
20	STATOR CAP
21	TERMINAL ASSEMBLY
22	DRAIN FLANGE

Figure 5. Typical canned motor cross-section.

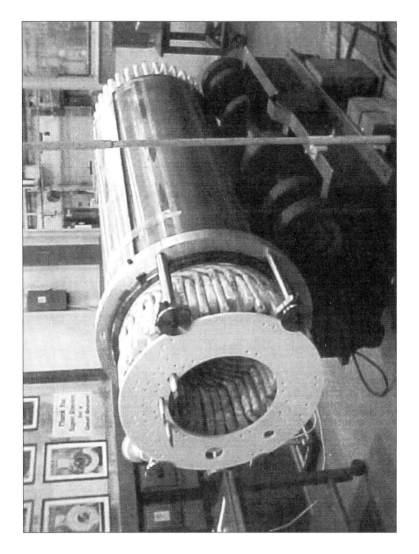

Figure 6. Fully connected motor winding. (21 Dec, 98)

Westinghouse Series 8058, Design Case T Motor Performance Characteristics at Approximately Constant Torque Output		
Running Performance in ISO 32 Oil at 60°C	**6760 vac 60 Hz**	**2253 vac 20 Hz**
Voltage/Hertz (V/Hz)	112.7	112.7
Speed (rpm)	1779	580
Power Factor (%)	74.7	74.3
Line Current (amps)	179	165
Input Kilowatts (kw)	1568	479
Output Horsepower (hp)	1700	567
Efficiency (%)	80.8	88.3
Kilo-Volt-Amp (kva)	2100	644

Figure 7. Calculated motor performance.

C556/012 © IMechE 1999

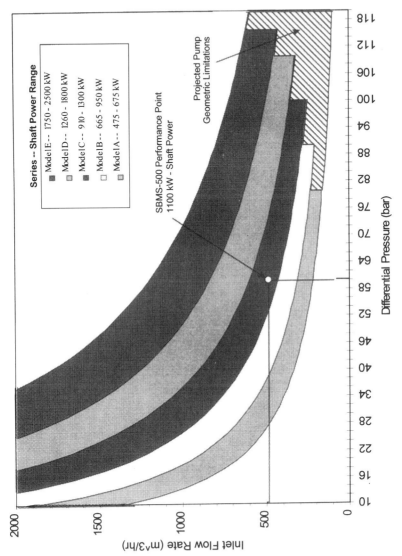

Figure 8. Preliminary family design curve.

C556/006/99

Applying multiphase screw pumps subsea

P COOPER
Ingersoll–Dresser Pump Company, New Jersey, USA
A J PRANG
Ingersoll–Dresser Pump Canada Inc, Ontario, Canada
P U THAMSEN
Ingersoll–Dresser Pumps Hamburg, Hamburg, Germany

Synopsis

Synchronised, twin-screw pumps have been proven in land-based installations as capable of handling gas void fractions of up to 100 percent whilst at the same time maintaining full pressure rise without a change in rotative speed. These pumps commonly range from 50 to 1,600 kW in power level. The reliability currently achieved by these pumps on the surface suggests deploying a package comprising a multiphase pump coupled hermetically to a submersible motor to boost the flow from one or more subsea wells. The required subsea engineering expertise resides in the same corporate family that manufactures both pump and motor.

1. INTRODUCTION

Multiphase boosting enables the economic recovery of oil and gas in any combination from remote locations, whether they be on the surface, on platforms, or subsea. As remote subsea wells deplete, the oil production can be quite small and so cannot be economically recovered unless correspondingly remote multiphase pumps are pressed into service.

There has been considerable interest in subsea multiphase boosting for several years; however, with the exception of one or two prototype installations, little or no subsea multiphase pumping has been done to date (1,2). While the economics of a successful subsea installation are attractive, questions remain about the reliability of the pump as it responds to the inevitable variations in flow of gas and liquid, system upsets, etc., of a subsea production system. Confidence has been building in the past few years as more and more multiphase pumping development and experience has been obtained in surface installations in oil fields, and this is providing a new impetus for subsea applications (3,4,5).

Gas-oil-water mixtures

Wells requiring boosting normally produce mixtures of gas, oil and water in proportions that can vary considerably with time at the pump inlet. Gas void fractions (GVF) of 0.95 or more (i.e., 95% gas by volume) are commonly encountered. GVF is related to the oft-quoted gas-oil ratio (GOR) or the mass of gas in standard cubic feet per barrel of oil (scf/bbl) through the volume flow rate fraction (GLR) of gas (Q_G) to liquid (Q_L):

$$GVF = GLR / (1 + GLR)............\ldots\ldots\ldots\ldots\ldots\ldots\ldots\ldots.1$$

where

$$GLR = (GOR) (T/T_{std}) (p_{std}/p) / (5.615 \text{ ft}^3/\text{bbl})....................2$$

and T is absolute temperature and p is pressure. Standard temperature and pressure are 15 C and 1.01325 bar Abs. (14.7 psia) respectively; so that $T_{std} = (273.15+15)$ K.

This mixture must be pumped to pressures p_2 as much as 50 bar or 700 psi above suction pressure p_1. Multiphase pumping tends toward an isothermal process – even at high GVF – as a) there is a considerable amount of liquid being circulated through the pump at all times, and b) the much greater heat capacity of the liquid in comparison to that of the gas produces a very small increase in the temperature of the liquid for each pass that it makes through the pump. In this case, the ideal power is

$$P_{isoth} = m R T_1 \ln (p_2/p_1) + Q_L \Delta p ....\ldots\ldots\ldots\ldots\ldots\ldots\ldots.3$$

where m is the mass flow rate, and R is the gas constant. Pumping efficiency is then this power divided by the shaft power – or, for overall efficiency – by the electrical input power.

Subsea multiphase pump deployment

The installation of a multiphase pump subsea involves subsea engineering that is considerably more expensive than the pump package itself (5.6). Yet, several advantages justify such an investment:

- Pump any GVF from zero to 1 (0 to 100 percent gas).
- Eliminate gas-oil separating plants (GOSP's) at remote locations
- Extend life of platform and wells
- Reduce wellhead pressure, increasing oil production.
- Increase step-out from platform or floating production, storage and offloading system (FPSO).
- Allow tie-back to deepwater wells from shelf platform.
- Assure flow via the higher pump pressure differential, the higher average velocity reducing wax and hydrate formation.
- Deploy pump(s) to serve one or more wells.

Types of multiphase pumps

Gas-liquid mixtures have been encountered for many years in various industrial processes, and this has led to developments in both 1) rotodynamic or impeller pumps (7) and in various kinds of positive displacement machines. In the latter category 2) rotary positive displacement screw pumps have emerged as dependable multiphase pumps (8). The distinguishing characteristics of these two types of pump are as follows (9):

1. Rotodynamic pumps operate as fluid dynamical machines, depending on speed and fluid density to develop pressure. Thus, to maintain pressure rise Δp at higher GVF, a higher pump speed is needed. The typical "helico-axial" configuration – comprised of many axial-flow stages – converts shaft torque into fluid angular momentum. Sudden changes in fluid density, as would occur in slugging, produce sudden changes in torque.

2. Rotary positive displacement pumps develop pressure hydrostatically – generally in a single stage – and so do not depend on the pump speed or fluid density. The inlet of the pump is walled off from the discharge – by the meshing of the screws in the popular two-screw configuration (Figure 1). The shaft power is simply the displacement volume rate Q_d times the pressure difference Δp across the pump; and the shaft torque is this power divided by the shaft angular speed ω. Thus, at constant speed and Δp, slugging has a relatively small effect on shaft torque, though the quantity of liquid pumped is reduced by the volume of gas that then takes up a larger portion of Q_d.

Screw pumps are the focus of this paper, because the small sizes needed for the low-flowing remote wells are relatively inexpensive. Further economies are to be had in that they can be driven subsea by correspondingly small, constant-speed, submersible electric motors; thereby eliminating the need for installing variable frequency drives (VFD's) subsea.

2. MULTIPHASE SCREW PUMP CONSTRUCTION

The basic multiphase screw pump design is similar to that of screw pumps proven on viscous liquid applications for many years (10). This design incorporates two parallel shafts with two screws on each shaft. The two screws mesh together in close fitting body bores to provide the positive displacement pumping action of both liquid and gas as illustrated in Figure 2. The two shafts are synchronised with timing gears to ensure proper meshing of the screws with no metal-to-metal contact of the screws. The screws on each shaft pump in opposite directions from the ends to the middle. This design doubles the capacity of the pump and provides balanced axial loads to minimize bearing loads.

Bearings, timing gears, seals, and rotors

For all multiphase fluid applications the normal external bearing and timing gear design is used (Figure 3). In this configuration, the timing gears and bearings are mounted outside of the main body, in separated chambers, and are lubricated by a separate lube oil system. Mechanical seals are normally used to seal the shafts – preventing fluid loss out of the pumping chamber.

The rotors are made of the integral design with the screws and shafts made from one piece to provide extra stiffness. The extra stiffness ensures that the rotors do not contact the body bores during high-pressure operation. This is essential in high-GVF applications, since there is no high viscosity fluid film to support the screws in the body bores. The tips of the rotors have a hard Stellite coating to resist wear and better tolerate solids such as sand.

Screw pump body and sealing liquid for high GVF

The standard screw pump body has two parallel bores, which house the screws and convey the product from the low-pressure suction area to the higher-pressure discharge flange. In multiphase applications, the body can include an integral separation chamber, which separates some of the liquid from the product being pumped. This liquid is then recirculated back to the pump inlet to lubricate, seal, and cool the screws in the bores at high gas void fractions. This provides cooling and lubricating liquid for the seals as well as sealing liquid for the clearances adjacent to the pumping screws. This feature allows the pump to operate at GVF values of 1 (i.e., 100% gas) for extended time periods, as long as there is some recirculating liquid available to provide the sealing and cooling required.

This design can be achieved with either a cast one-piece body or it can be made in a cartridge-type design with a replaceable inner bore section and an outer shell. This arrangement allows for the use of a shear ring, which simplifies assembly and disassembly. The bores of the pump are coated with industrial hard chrome with a hardness of 65-70 Rc to resist wear from solids such as sand which are commonly encountered with multiphase mixtures.

3. FIELD EXPERIENCE WITH MULTIPHASE SCREW PUMPS

A number of rotary, synchronised, twin-screw pumps have been used in applications involving multiphase products over the last 30 years. Multiphase, geared, screw pumps have been used in the chemical processing, pulp and paper and petrochemical industries (8). In the last decade, the multiphase pumping applications have concentrated on petroleum products, specifically oil wells. Most of these applications are surface located and generally onshore. Several are located in a remote area of Alberta, in western Canada. The range of typical pump sizes, illustrated in Figure 4, covers intake flow rates to 19,000 m^3/d (120,000 bpd) at a Δp of 50 bar (725 psi) (higher flow rates at lower Δp) and power levels to 1.6 MW.

Small pump applications

These applications involve a relatively small multiphase screw pump, model MP1-125, which is shown in Figure 5. In one application, this multiphase pump is connected to a field of approximately 50 small oil wells. The pump was designed to operate at a GVF of 0.663. This and the other design conditions of service are given in Table 1. This pump provides for ingestion of 983 m^3/d of gas together with a total liquid capacity of 500 m3/d or 3145 bpd at a pressure rise of 20.7 bar (300 psi).

This pump is equipped with a special cast body with an integral liquid separating chamber. As the multiphase mixture exits the screw area it must pass though the separating chamber where the fluid velocity is reduced, thus allowing the liquid component to separate from the gas and settle into a chamber under the screw bores (Figure 3). This liquid is then recirculated through cyclone separators and fed back to the inlet areas by way of the mechanical seals. This provides cooling and lubricating liquid for the seals as well as sealing liquid for the pumping screws. This feature allows the pump to operate at GVF values of 1 (i.e., 100% gas) for extended time periods, as long as there is some recirculating liquid available to provide the sealing and cooling required.

This pump has been in service since late 1996 and has provided significant benefits to the customer. There were some initial problems with leakage of the mechanical seals, which are flushed by the product being pumped, the flush fluid first being cleaned in cyclone separators. The high solubility of the gas in the liquid component reduced the effectiveness of the separator system when compared to the air/water multiphase product used for the in-house testing. A change was made to the seal design to provide a more tolerant seal arrangement. The revised seal design utilized cartridge mounted rubber bellows seals with tungsten carbide rotating and stationary faces. This change in seal design, along with the installation of dual cyclone separators at each end of the pump significantly increased the operating reliability of this pump. Similar problems involving mechanical seals have been encountered and solved by virtually all manufacturers of multiphase pumps (11).

A new, computerized monitoring system called PumpTrac was also installed on this pump to allow remote monitoring of the various operating characteristics. This system allows the customer or the factory to monitor such variables as vibration, bearing temperatures, rotating speed, inlet pressure, outlet pressure, and motor amperage from remote locations via telephone lines.

The customer has advised that this multiphase pump has provided significant benefits for his production installation. The pump is installed downstream of the free water knockout tank, and the speed is controlled to reduce the pressure in this tank from the original 13.8 bar (200 psi) to 5.2 bar (75 psi) gauge. With the present wells, the tank pressure is even lower [corresponding to a pump inlet pressure of 2.8 bar (40 psi) gauge] and is achieved with the pump operating at only 60-90% of full speed. The actual GVF of the product varies from 0.75 to 0.90. The benefits include an 8% increase in oil production with no increase in total power draw of all pumps at the site and a reduction in the system pressure upstream of the pump. Another significant benefit is the reduced Δp of the small downhole progressive cavity pumps located in each well. This has significantly reduced the wear, and they are experiencing approximately twice the normal life on these pumps. This in turn has significantly reduced the maintenance costs involved with pulling the pumps from the wells when service is required.

As a result of this successful application, two similar MP1-125 multiphase pumps have been sold to another customer in western Canada and have been operating since early 1998. In this application, the pumps are sized for higher capacities by using a larger screw pitch and operating at faster speeds, up to 3000 RPM. These pumps were designed to operate at a GVF of 0.895, with an inlet gas volume of 1775 m^3/d, oil capacity of 134 m^3/d and water capacity of 75 m^3/d. Variable speed drives allow these pumps to be adjusted for widely ranging volume conditions. Design pressures are 6 bar inlet and 26 bar discharge. These design conditions of service are given in Table 1.

These pumps draw product from five wells each in a remote oil field. The experience with the modified seal and flush arrangement with the original MP1-125 pump was used to minimize the seal problems with these pumps. In spite of this, seal leakage is the most frequent problem experienced with these pumps; but minor design changes have improved the reliability, and the benefits of the multiphase pump have been realized.

Large pump applications

Another installation involves a much larger multiphase screw pump, mounted on an offshore oil platform in the Middle East. This installation includes an MP1-275 size multiphase screw pump, packaged complete with the VFD system, and all of the required lube oil and seal flush systems, monitors and controls. This application involves many varying flow conditions. The pump was designed for an inlet GVF of 0.942, with an inlet gas volume of 6831 m^3/d, oil capacity of 297 m^3/d and water capacity of 127 m^3/d. The design pressures (absolute) are 7.9 bar inlet and 49.3 bar discharge. These design conditions of service also are given in Table 1.

4. SUBSEA MULTIPHASE PUMPING UNITS

For subsea applications the standard screw pump design must be modified to allow submerged operation at high ambient pressures. Similarly designed, submersible, liquid-cooled motors exist (Figure 6). These are supplied by the same manufacturer as the pumps, and they successfully operate subsea in a variety of tasks (5). Figure 7 shows a multiphase, synchronised, two-screw pump packaged together with one of these motors.

Design of large subsea pump
The MP1-275 size listed in Table 1, when reconfigured in this manner for subsea service, is conservatively resized to ingest 5556 m^3/d (34,950 bpd) of liquid and gas and to increase the pressure by 44 bar. At a nominal GVF of 0.46, the unit consumes 525 kW. The design incorporates a high-pressure fabricated screw pump body with a replaceable cast liner. This design provides an integral liquid separator, which separates the liquid and provides a reservoir at the lower area of the body to store the separated liquid. This separated liquid is recirculated back into the suction areas of the screws to provide the required sealing liquid at very high GVF's.

The cast liner portion of the body contains the precision ground bores where the screws operate with controlled clearances. O-ring type sealing joints are employed between the liner and the body, which are suitable for pressures up to approximately 138 bar (2,000 psi). For applications above this pressure, different gasketted joint designs are possible to permit this pump to handle high differential static pressures, which could be encountered in a deep subsea application.

Seals and oil-supplied components
This design utilizes two mechanical seals at the drive-end of the pump to seal the product from the lube oil cavity. The seals operate in the lube oil, which provides lubrication and cooling. The drive-end mounted timing gears and thrust bearings are also mounted in the same lube oil chamber, which is also connected to the submersible oil-filled motor. The wall sections and sealing joints are presently designed for the 138 bar operating pressure but can be redesigned to handle higher pressures for deeper well applications.

Compensator
A differential-pressure compensator is connected between the pump inlet and the lube oil chamber to control the differential operating pressure on the mechanical seals. The differential pressure compensator shown in Figure 7 utilizes an internal piston mechanism to regulate the differential pressure across the seals to 10% of the pump suction pressure to ensure there is no leakage of process fluid into the bearing, timing gear and motor housing

cavities. Other types of pressure compensators can also be used to maintain a constant differential pressure across the seals. The compensator also provides a reservoir of lube oil to make up for minor seal leakage. The sizing and type of compensator are dependent on the seal design and operating conditions and would be sized to provide adequate seal life in subsea applications.

Product-lubricated bearings and monitoring

The line bearings at the nondrive-end of the pump are designed as product-lubricated sleeve bearings. These silicon carbide bearings are capable of supporting the shaft loads and operating in the liquid available. A separate flush porting arrangement, not shown, directs the separated liquid in the reservoir to the ends of these bearings to provide suitable lubrication.

Temperature monitors can be installed to shut down the pump in the event of high seal or bearing temperatures. This elevated temperature condition would occur only if a slug of 100% gas, of significant duration, were pumped. Under such conditions (GVF = 1), the supply of recirculating liquid eventually depletes – being carried off by the gas and leaving insufficient lubrication and cooling for the bearings. Similarly, if the clean lube oil supply is lost, high seal temperature provides a warning for shutdown prior to failure. The pump can be readily restarted when some cooling liquid is available from the product.

5. ADVANCED SEALLESS DESIGN

Recognizing that the mechanical seals are the weak point in the pump design, an alternate design approach is possible to eliminate these seals. With advanced materials available, all the bearings and timing gears can be product lubricated. The resulting pumping unit will be similar to that of Figure 7, except that the timing gears and thrust bearings would now be placed at the nondrive-end with the nondrive-end product-lubricated bearings, as shown in Figure 3 – but with pumped product being supplied there instead of oil. The use of materials or coatings such as tungsten carbide would allow product lubrication of these components. The pump body would be the same configuration as described above with the integral liquid separation chamber. Liquid from this chamber would be further purified with cyclone separators and then directed to the bearing and timing gear chambers to provide adequate lubrication for these components.

The nondrive-end mounted thrust bearings would be equipped with trapped axial faces to provide thrust control in either direction. While the hydraulic loading of the screw pump is totally balanced in the axial direction, accurate axial positioning of the shafts in relation to the timing gears is important to keep the screws synchronised and ensure that they do not contact each other. The timing gears and thrust bearings are located at the nondrive-end of the pump to facilitate design and assembly with the product-lubricated bearings.

In this case, liquid product would also cool a canned version of the motor. This same liquid, separated from the multiphase product, is circulated through the gear and bearing housings. Temperature sensors and shutdown controls would be used to shut down the pump and motor in case the supply of liquid is not sufficient to lubricate the bearings, and timing gears; and to cool the motor. An external cooling coil system can also be included with the system to provide additional cooling for the lubrication fluid.

This design provides significant advantages by totally eliminating mechanical seals. This sealless pump requires some development in the areas of product-lubricated timing gears and bearings but there are good success examples with new materials, suitable for these services. The potential benefits of the sealless pump in this environment make this development viable.

6. SUBSEA ENGINEERING

The success of any subsea multiphase pump project depends on the unique capabilities of an experienced subsea engineering team. When this team is part of the same organization as the supplier of the pumps and motors – as is the case for the subsea multiphase work being described here, the economies and efficiency of this unified approach are evident and can contribute significantly to the success of the project. Subsea engineering in this case involves the following elements:

- Submersible pump-and-motor unit
- Compensator system
- Mounting structure for the pump-and-motor unit
- Installation and maintenance
- Control umbilical
- Safety and certification
- Process flow
- Electrical system

Submersible motors (The pump has been described above.)
Submersible motor technology has proven reliable in many difficult applications, including several installed subsea. These motors can readily be utilized to drive multiphase screw pumps. A range of subsea submersible motors varying in power from 1 kW to 5 MW and in speed from 200 to 3,500 rpm – at voltages up to 10 kV – is available from the same manufacturer as supplies the pumps. (See Figure 6.) The motors are of the 3-phase asynchronous AC cage type, operating at input voltages of at least 2.6 kV. The motor is filled with oil and is housed in a pressure-proof casing, suitable for the shut-in pressure of the well. The housing is coupled directly to the pump such that the oil that surrounds the motor also lubricates the seals and timing gears of the pump. This oil also provides cooling to the motor via an integral set of heat exchanger tubes fitted to the motor casing.

The electrical connections to the motor are made by means of two pressure-proof penetrator assemblies, one for the high power conductors for the motor windings, and the other for the instrumentation sensors (motor temperature and water ingress). These penetrators are based on a down-hole packer penetrator design that is widely used for electrical submersible down-hole pumps and is qualified for pressure barrier use.

Compensator system

As described above, the pressure compensating oil inside the motor housing is stored in a separate compensator unit, which is directly piped to the motor housing. The compensator is a piston-type, with one side of the piston connected to the inlet area of the pump and the other to the motor housing. The compensator has a large spring, which provides the over-pressurization of the oil to the motor and is fitted with a sensor for detecting the piston position and therefore the remaining volume of compensating oil.

Compensators can be recharged by a Remotely Operated Vehicle (ROV) via a set of valves on the ROV intervention panel on the manifold, or conversely a supply of compensating oil may be provided via a hose in the control system umbilical. In either case, the oil supply will ensure that seal leakage is accommodated.

The leakage rates of the mechanical seals will affect the sizing of the compensators. Very accurate seal faces can be used with high contact pressures to minimize the leakage rates, but this will result in lower seal life. A compromise seal design is recommended to provide adequate seal life without excessive leakage of the oil into the product. The compensator is sized to allow storage of typically one year's supply of oil.

Mounting structure for the pump-and-motor unit

The design and type of mounting structures used can vary significantly, depending on the size of the equipment, the depth of the installation and the nature of the seabed in the area. In some cases a concrete pile design is used to anchor the structure to the seabed. If the equipment is of sufficient size and the environment suitable, the equipment van be mounted on a "gravity base" structure, using its own weight and size to retain its position on the seabed.

The mounting structure for the multiphase pump-and-motor unit shown in Figure 8 is of the concrete pile design – to be anchored to the seabed. This structure comprises two main components, the Pump/Motor Module (PMM) and the Pump Module Landing Base (PMLB). The PMLB can contain a series manifold system and suitable valves and controls to permit operation and isolation of the PMM. The PMLB is installed on a pile (or a gravity base) and the inlet and outlet jumpers or flow lines connected to two connector/valve block assemblies that allow the PMM to be disconnected and removed to the surface for maintenance whilst leaving the PMLB and flow lines intact. The flow lines can be either rigid steel or flexible, and their connection system chosen to suit field design and water depth.

Installation and maintenance

The design of the mounting structure and connections is such as to facilitate installation and maintenance. Applications for a multiphase pumping system such as this will often be in deepwater, so diverless and guidelineless methods have been chosen for all intervention and installation / retrieval operations. (This does not however preclude the use of divers in shallow water applications). The mounting structure or package is designed to be installed in four stages:

1. The pile is established.
2. The PMM and PMLB are lowered together from the surface and landed and locked on the pile.
3. The jumpers that connect the Pump Package to the wellhead and the export flow lines are installed.
4. The control umbilical is laid away from the Pump Package towards the host platform where it is terminated.

As the PMM will be lowered from the installation vessel on a single lift wire during subsequent installations, a conical guidance system is used to locate the PMM on the PMLB, and the PMLB on the pile. A simple running tool carries the PMM (or the PMM mated to the PMLB) during these operations. The package design is kept simple and compact to allow installation from a monohull vessel. These operations all depend on interfaces between the various components that make for easy and quick connection via an ROV (or diver). Figure 9 is an illustration of typical interfaces for such connections.

Control umbilical

The control umbilical connects the Pump Package to the host platform and provides power and control services to the system. This umbilical terminates close to the PMLB and is connected to the PMM by two ROV installed jumpers.

Safety and certification

In order to satisfy the certification requirements of the flow line, all components in the Pump Package that are exposed to process products must be rated for the operating pressure of the flow line. This includes the pump and motor housings, pressure compensation system and all process pipework and valves. As operating conditions and well shut-in pressures vary with each application, a definitive pressure rating for the system is not currently defined. However, typical shut in pressures that could be catered for would be in the region of 350 bar.

Process flow

An ROV operated bypass valve connects the inlet and outlet flow lines on the PMLB, which allows non-pumped flow through the system when the Pump is not installed. (See Figure 10.) On either side of this bypass valve, two ROV operated isolation valves direct the flow to and from the PMM and allow operation of the flow line with the PMM removed (after the installation of pressure caps on the connector mandrels in place of the PMM).

After Installation, flow line jumpers are connected to vertical connector mandrels on the PMLB. In turn, these mandrels are piped to flow line connector/valve blocks that make the process connection to the PMM and house the bypass and isolation gate valves. The PMLB to PMM flow line connectors are operated hydraulically by the ROV during installation and retrieval operations. An ROV of the type typically used for these operations is illustrated in Figure 11, and the completed installation is shown in Figure 12.

On the PMM a further pair of isolation valves are fitted on either side of the multiphase pump which are used during installation and retrieval operations. An actuated bypass valve is fitted across the inlet and outlet of the pump, which allows the flow through the pump to be controlled by the Pump Package control system. This bypass valve will remain in its open position until the pump is operating at full speed, whereupon it will be closed to bring the pump on-line.

As the pump is a fixed displacement type, if it stops rotating then the flow will also stop. For this reason the bypass valve is a fail-safe open type, having a spring actuator that will drive it to the fully open position in the event of a failure of the pump or control system. If deemed necessary, flushing and vent valves allow the connection of the workover flushing system to the pump prior to retrieval via the running tool.

Electrical System

The motor input voltage (of at least 2600V) is supplied directly from the host platform via a 3-core power cable incorporated in the composite control umbilical. The motor operates at a fixed frequency and relative shaft speed (based on load and slip values).

As indicated in Figure 10, electrical power is supplied from the host platform, where the necessary switchgear and electrical control equipment is located. This equipment comprises MV switchgear, a step-up transformer, monitoring equipment and protection circuits.

At the pump end, the umbilical termination is landed close to the PMLB. The termination has a stab-plate connection system that allows two ROV installable electro-hydraulic jumpers to make the connection between the termination and the PMM.

On the PMM, the high power cable from the static stab-plate connects to a pressure resisting bulkhead penetrator in the motor housing. This penetrator allows the well shut-in pressure to be contained inside the motor housing.

7. CONCLUSIONS

Multiphase boosting enables several economies in oil production, all of which have driven the application of multiphase pumps in numerous surface applications. These drivers are the ability to

- pump from new or existing wells to platform or FPSO
- make remote, marginal wells productive
- shorten depletion times, saving operating costs

Rotary two-screw pumps have demonstrated the feasibility and flexibility required for the varying and severe conditions of the high-gas void fraction (GVF) wells typically involved. The development that has occurred as a result of this experience has increased multiphase pump reliability to the point that these machines are now robust, low maintenance packages that can be installed subsea.

Multiphase pumping using two-screw pumps is a proven technology. They are available in power ranges from a few kW to 1.6 MW and can operate at constant speed with GVF varying from zero to 1, at the same time producing low torque variations with such changes in process parameters.

In addition to pumps, field proven equipment for deepwater application is available:

- Electric motors
- Seabed anchoring
- Retrievable architecture
- Remote flow line and umbilical connection systems
- Electro-hydraulic controls
- ROV intervention

This equipment, together with the subsea engineering required to design, procure, deploy and maintain a complete subsea multiphase pumping unit or station are all available from the same organization, allowing for an efficient, integrated engineering approach.

ACKNOWLEDGEMENTS

The subsea engineering material in this paper was furnished by Messrs. John Mair and Gareth Kerr, who were in the employ of SubSea Offshore, Ltd. (SSOL), at the time of writing. They are currently employed by Det Sondonfjelds–Norske DampskibsselskAB (DSND). Others at

SSOL, a Halliburton company, also were involved in this effort. Ingersoll-Dresser Pump Company is a joint venture of Ingersoll-Rand Company and The Halliburton Company.

REFERENCES

1. Assayag, M., et al: "Subsea Boosting Systems: Advances, Trends and Challenges," World Oil, November 1997, pp. 61-69.
2. Grenato, M.: "Field Operation of a Subsea Multiphase Boosting System," *6th International Conference on Multiphase Flow in Industrial Plants*, Milan, Italy, September 1998.
3. Neumann, W.: "Efficient Multiphase Pump Station for Onshore Application and Prospects for Offshore Application," *Proceedings of the Eighth International Pump Users Symposium*, Texas A&M University, March 1991, pp. 43-48.
4. Caetano, E. F., et al: "SBMS-500: Cooperation on a Multiphase Flow Pumping System," Offshore Technology Conference Proceedings, Paper 8454, 1997.
5. Cooper, P., et al: "A Versatile, Multiphase Two-Screw Pump Package for Subsea Deployment," *Offshore Technology Conference Proceedings*, Paper 8861, May 1998.
6. Elde, J., et al: "Economics in and Practical Use of Multiphase Booster Pumps," *Offshore Technology Conference Proceedings*, Paper 8860, May 1998.
7. Schiavello, B.: "Rotodynamic Pump Performance With Liquid-Gas Medium as Related to Fluid Conditions and Hydraulic Design," *6th International Conference on Multiphase Flow in Industrial Plants*, Milan, Italy, September 1998.
8. Prang, A.: "Rotary screw pumps for multiphase products," World Pumps, November 1991, pp. 18-23.
9. Cooper, P., et al: "Tutorial on Multiphase Gas-Liquid Pumping," *Proceedings of the Thirteenth International Pump Users Symposium*, Texas A&M University, March 1996, pp. 159-173.
10. Prang, A.: Selecting Multiphase Pumps," Chemical Engineering, February 1997, pp. 74-79.
11. "Applied Multiphase Pump and Multiphase Meter Technology," an SPE Applied Technology Workshop, 8-9 May 1997, Houston, Society of Petroleum Engineers, Richardson, Texas.

Table 1. Multiphase screw pump applications

Design Condition of Service	Units	MP1-125 – Surface Western Canada – A	MP1-125 – Surface Western Canada – B	MP1-275 – Platform Middle East
Disch. Pressure	bar (psi) gauge	25.8 (375)	26.0 (377)	48.3 (700)
Inlet Pressure	bar (psi) gauge	5.2 (75)	6.0 (87)	6.9 (100)
Liquid Capacity	m³/d (bpd)	500 (3,145)	209 (1,315)	424 (2,667)
Oil	m³/d	425 (85%)	134 (64%)	297 (70%)
Water	m³/d	75 (15%)	75 (36%)	127 (30%)
Total Inlet Vol.	m³/d	1,483	1,984	7,255
GVF at inlet		0.663	0.895	0.942

Figure 1. Meshing of the screws in a screw pump. Pumping screws form chambers that convey fluid axially from the inlet to the discharge as the screws rotate.

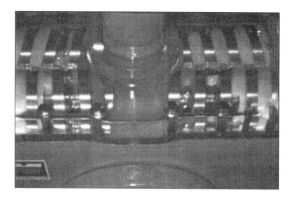

Figure 2. Multiphase fluid in a two-screw pump. Oil and gas froth is conveyed axially from both ends to the center discharge port. Cavities of gas and liquid droplets can also be seen.

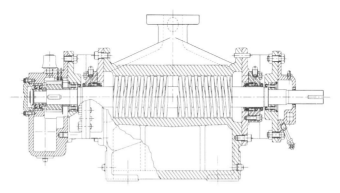

Figure 3. Multiphase two-screw pump. Liquid reservoir is shown at the bottom.

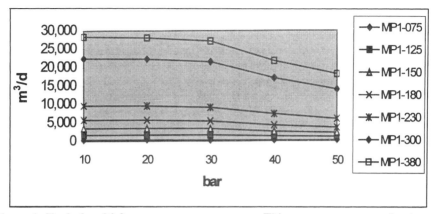

Figure 4. Typical multiphase screw pump coverage. This range covers power levels up to 1.6 MW and pressure-rise values to 50 bar. Volumes shown are total at pump inlet.

Figure 5. Surface Installation of Multiphase Screw Pump. Located in western Canada, this "MP1-125" pump takes in 1500 m³/day (9400 bpd) of gas and liquid.

Figure 6. Submersible motor for subsea pump. This 220 kW, 3,500 rpm unit is one of a range from 1 kW to 5.5MW.

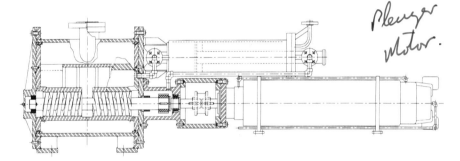

Pleuger Motor. (handwritten)

Figure 7. Subsea multiphase screw pump unit. Totally enclosed, pressure-containing environmental envelope contains pump, submersible motor, coupling, and pressure compensator. Compensator maintains pressure in clean oil that fills motor and timing gears at level slightly above pumped-fluid suction pressure.

Figure 8. Mounting structure for multiphase pump-and-motor unit. Includes pump/motor module (PMM) and pump module landing base (PMLB). Pump and motor can be seen in the middle of the PMM, with the compensator above. The lower platform is the topmost part of the PMLB, which is permanently mounted on the sea bottom.

Figure 9. Example of interfaces for subsea connections. These interfaces allow for piping and electrical connections to be made via a remotely operated vehicle (ROV).

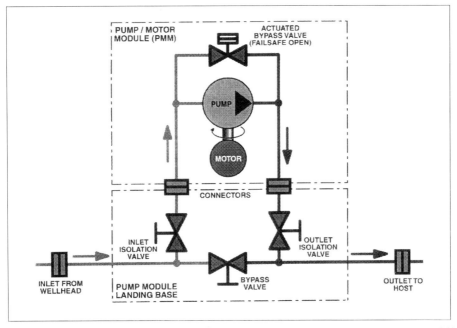

Figure 10. Process flow diagram. PMM can be removed for maintenance. The ROV that does this also valves off the line coming from the wellhead to the pump and opens the bypass valve so that product can flow in the absence of the pumping unit.

Figure 11. Work-class ROV. This robotic unit is controlled from a floating vessel and performs construction and main-tenance tasks.

Figure 12. Subsea installation of multiphase pumping unit. Jumpers are shown connecting the wellhead to the pump intake and the pump discharge to the platform manifold. The darker-shaded pipe contains control umbilical lines from the platform or FPSO.

C556/015/99

Low-flow, high-head – normal speed pumps

L W GATEHOUSE
Process Systems International Limited, Northampton, UK
J D HOOLEY
Pumping and Technical Services, Macclesfield, UK

SYNOPSIS

The increasing industry need for low flow, high head applications with centrifugal pumps, using normal synchronous speeds, is re-examined. The co-authors, each with over 30 years of industry application experience, have taken a fresh look at these requirements, using modern 'milled-from-solid' manufacturing techniques.

Introduction and background

The API oil, gas and process market has, over the last decade, increasingly required low flow, high head pumps for a number of applications, as processes have more closely refined the quantity and pressure requirements.

The conventional ways of achieving these more stringent low flow, high head duty parameters have included positive displacement, high speed and multi-stage centrifugal pumps. Some such units have drawbacks either from an operational flexibility, cost, higher maintenance or environmental noise viewpoint. Figure 1.

The co-authors have experience of low flow, high head centrifugal pumps dating back to the early 1970's period. Using the old form of designation, sizes such as 2" x 2" x 16" and 3" x 3" x 16" pumps were produced successfully for Shell, Stanlow and BP, Baglan Bay, the numbers indicating suction, delivery and impeller diameters. Figure 2.

These early units had only a couple of snags, one due to casting limitations and the other of noise caused by the impeller blade passing frequency in the casing cut-water areas, the blade numbers causing a harmonic 'siren' affect.

The co-authors, using their associated companies, decided to take a fresh look at these applications, but utilising modern manufacturing techniques.

The ability to use a centrifugal pump has significant advantages for operators, which can be summarised primarily as:

Operational flexibility over a wide performance curve range.
Simplistic design.
Lower maintenance costs.
Moderate capital costs.
Durability.

Our first approach was predicated by a requirement from Amec Process & Energy Limited for an application for the Mobil (now BP) Coryton Refinery.

The main operational parameters were:

Q	=	$6.65m^3/hr$
H	=	223m
Npsha	=	2.8m
Fluid	=	Hydrocarbons + 3% SO^2
T	=	360°C max.
Materials	=	API grade A8

To be capable of hot standby service.

Basic design

Our policy of avoiding the use of castings for pump construction, which was formulated twenty years ago, followed bitter experience with defective supplies and has resulted in a level of expertise in making casings and impellers from solid billet/fabrications which we feel is probably unique.

The substantive nature of components milled from solid provides a strength and robustness, without any great capital cost penalty. It also allows us to individually engineer small volumes of tailor made pumps without recourse to pattern equipment and pattern modifications. An added benefit of 'milled from solid' construction is that wrought materials tend to have a denser micro-porous structure and practical experience over the years has shown such construction to be 4 to 6 times tougher than cast equivalents. An associated side benefit of this method is the ability to be more resistant to corrosive attack. Figure 3.

Impeller design

The increasing sophistication of N.C. Milling equipment enables manufacture of impellers which are exactly as designed - this not being the case with the majority of cast impellers due to core shift, etc. A valuable aspect of these techniques is that all the internal surfaces are in a machined condition, infinitely better than most cast surfaces in terms of surface friction.

This element of the design of large diameter high head impellers, assumes almost alarming significance in terms of power absorbed (and therefore operational efficiency). To give an example from recent practical research, a plain machined disc of 400mm diameter rotating at 3550 rpm in an "open" casing, absorbed approximately 38kw to do very little hydraulic work. It was of interest to observe that this disc generated a high flow rate, some $170m^3$/hr, but at a low pressure - 2 PSI(g).

Some study of the work done by Tesla at the beginning of the century on plain disc pumps is of interest, as also is Gibson and Unwins research, together with Stepanoff's development of this, but it has to be said that none of this work seems to have been taken to an extreme of development. For example Stepanoff's results were indicating what he describes as a "decrease in the co-efficient of friction" as the casing walls more closely approach the disc - this would appear to be an incorrect use of the terminology, the actual observed effect was that the power absorbed reduced, albeit slightly, as the disc/cavity clearance decreased.

Bursting stresses in these large diameter impellers are obviously a major consideration and in our largest 510mm diameter version, would not be acceptable to the writers in cast construction. However whilst the stresses in a disc rotating in air are easily calculable, we have mused over the possibility that an actual impeller in liquid may well be subjected to forces opposing the centrifugal stress, due to the pressure in the waterway, which in the case of a circular rather than volute shape, is substantially constant all around, a point which we have verified by actual observation during many performance tests.

A major feature of the design of low flow impellers, particularly where reasonable to low NPSHR figures are required, is to preserve the liquid velocity through the impeller within certain parameters. Using machined impellers it is practicable to look at the individual passages rather than the blades - that is to say, look at the hole, not the metal. Indeed, our lowest flow impellers do not have blades in the accepted sense, but rather a set of passageways milled from a solid disc. In a recent example of this construction of a small scale unit, the particular impeller in unshrouded form, gave 0.26m NPSHR at $0.9m^3$/hr flow at 3550 rpm and was the prototype for a resulting order for 17 such units in Hastelloy for vacuum extraction service at 150°C. (Shop tested at 150°C). Figure 4.

For low flow applications, to enhance the low NPSHR requirements, the impellers tend to have large eye diameters, resulting in lower inlet tip speeds. For say, a $10m^3$/hr flow rate, conventionally a $1^1/2"$ or 2" inlet is common, whereas a 3" inlet of a C.L. Pump has an inlet velocity of 0.66m/s, resulting in better performance parameters.

The subject of wear ring/balance hole relationships has been touched upon earlier. It is worth noting that careful attention in this area was rewarded on the Coryton pumps, with an increase in generated "pressure" of some 1.5 bar in a total of 23/24 bar, without creating any undue pressures within the seal chamber and without power penalty.

Another factor that had to be considered and contained was the seal area containment pressure generated by the impeller. This is overcome by selective component engineering, including rear neck rings (or none) and size, shape and angle of impeller balance holes. For high suction pressure applications, to 100 bar (g), front impeller breakdown restriction bushes can be used to contain fluid flow and forces.

Casing design aspects

It could be argued that a circular casing is only chosen to make manufacturing easy and eliminate pattern costs, both of which it does. However, apart from our aversion to castings, it was a conscious design decision, in spite of Stepanoff's dismission of circular casings. Our reasoning was that virtually no single stage centrifugal pump is going to achieve a good efficiency at what is regarded as low flow, i.e. Below $10m^3$/hr, so the benefits of a volute can be exacerbated by the velocity energy conversion tendency to de-stabilise the curve at flow rate readings of 0, 1, 2, 3, 4 and $5m^3$/hr.

This aspect also dictated our preference for a tangential outlet pipe into which we can fit different size venturi type nozzles to 'fine tune' the shape of the curve i.e. we can easily control velocity energy conversion. Figure 5.

The circular casing design also moves the minimum radial load towards low flow instead of best efficiency point and leads into the subject of unbalanced hydraulic radial loading. Our calculations for this are always based on the accepted criteria (C.S.A. x generated pressure x K) to establish bearing loading, but we have some reservations about the accuracy of this, as we believe the forces involved to be far less.

Our proposal to Amec P&E Limited was for two 80 x 40-400 CL API 610 pumps, as shown in sectional drawing number 1109a. Figure 6 refers. The hydraulic end was constructed in 'milled from solid' AISI 304L. The impeller and blades being form-milled from a solid billet so as to eliminate weld areas, the front shroud being screw fastened and tack welded for security, to the blade faces. In this way, the peripheral speed and hoop stresses could be safely contained with a large safety margin. This method has been successful in service on impellers up to 510mm diameter, where the peripheral speed approaches 96m/s @ 3500 rpm. For low flows of 1.0 to $5.0m^3$/hr, impeller tip widths between the shrouds of as little as 3mm are possible, using this construction method. Figures 7 and 8.

Initially the unit was designed with double back to back 7 series thrust bearings as per API 610. However both Amec and Coryton had some reservations regarding lack of bearing loading at low flows and low suction pressures, as bearing skidding has been known to cause premature failures. We concurred with this practical operational experience and changed to single deep groove angular contact bearings, which have stood the test of time at Coryton.

To illustrate the minimum flow capability and durability of the design, it was also required to be capable of a minimum design stable flow of 2.0m³/hr at a normal pumping temperature of 329°C. In reality the unit was required to operate below minimum design flow, albeit at a lower temperature, but at a much higher fluid viscosity. The pump operated under these conditions at 0.5m³/hr for a period of 48 hours, pumping out a column following an unscheduled process shutdown. The pump, when removed from service and examined, exhibited no detrimental effects to rotating or stationary components. Seals were replaced and the pump unit put back into service. Figure 9.

Seal area design

As mechanical seal requirements and hence life, are as critical on low flow, high head applications, as on much bigger units, we tend to design cone type seal housings with vortex breakers, to give the hard working seal the best possible environment.

Other special features

The Coryton application had two additional aspects that had to be addressed. One was that the fluid was temperature/viscosity sensitive. This was solved by utilising steam jacket areas for the casing backplate and seal housing, so the operator could control the viscosity using medium pressure steam and also to ensure the hot standby unit would always be available for service.

The other was the fluid temperature affecting the front radial bearing, due to heat transference along the shaft, in hot standby mode. Bearing manufacturers like to aim for a temperature limitation of about 110°C for their bearings. Our experience has shown that the conduction loss approximates 25°C per 25mm of shaft length. Therefore with process fluid at up to 360°C and M.P. Steam at the seal, the radial bearing temperature was predicted to exceed 110°C by a considerable margin. This aspect was solved by using refinery process water to cool a double cartridge water seal, located outboard of the radial bearing. We constructed this seal using a sleeve with a series of holes, to allow the process water to pass through and thus cool the shaft in this area, when the pump was stationary.

It can be argued that the design features highlighted in this paper are exceptions to accepted API 610 practice - but these are usually driven by user preferences, once they are aware that the design flexibility exists for small quantities of pumps. Figure 10.

We can and have, manufactured more API 610 compliant units, but for low flow, high head applications it is our experience that our flexible design approach has more advantages than disadvantages to our clients.

As the subject range of API pumps is custom engineered in small quantities, it is possible to design for critical applications that are specific to clients process conditions. Examples of this are indicated by the ability to tailor the curve shape to suit the system. Within reason, this can be 'flattened' or made more steep by the ability to machine the blades from solid using 3D milling techniques. This design manufacturing flexibility enables the pump range to achieve relatively low NPSHR figures commonly needed for low flow, high head applications, where available NPSH is limited. NPSHR figures in the region of 1-3m are commonly achieved, with some units achieving figures of under 1.0m NPSHR, depending on impeller style (open or closed) and rpm. Figure 11.

A final point on the technical side regarding low flow, high head units.. With properly configured tangential outlet nozzles, as opposed to the more usual centre line discharge favoured by the industry, many process applications can be achieved by varying aperture sizes, divergence angles etc. to "engineer" a pump to a given duty and not just by "orificing". A centrifugal pump is a velocity energy conversion device and the precise relationship between the impeller peripheral velocity, the velocity of the liquid within the waterway and the liquid velocity at the discharge flange are complex and readily susceptible to informed experimentation. It is a common misconception that centre-line discharge pumps are totally self venting - an examination of most designs often reveals that an air pocket can still exist just after the cutwater tongue.

With ultra-low specific speed machines such as the subject pump, a high velocity jet issues from the core of the casing throat and this is what we try to convert into pressure. A centre-line discharge inevitably involves a distortion of the perfect venturi shape we want to achieve, having unpredictable affects on the liquid flow. With a tangential discharge this jet can be accommodated within the tangential cone itself at low flows, i.e. Small hole sizes. Figure 12.

Centre-line discharges do of course have other advantages such as transmission of pipe loads to the bedplate, but in the sizing of the subject pump range the API 610 nozzle loads are easily achievable. Figure 13.

In conclusion we would put to the industry that, whilst API 610 is an acceptable engineering standard, it is our experience that some aspects are inappropriate to low flow, high head units. We prefer to regard API 610 as a design philosophy guideline - the summation of years of practical user experience. However, as with all engineering, we would caution against regarding it as a rigid dogma - if engineers do not question things, think laterally, try out new ideas and even re-apply old ideas, then how can we hope to progress.

© 1998   l. W. Gatehouse – P.S.I. Limited
         J. D. Hooley - Pumping & Technical Services

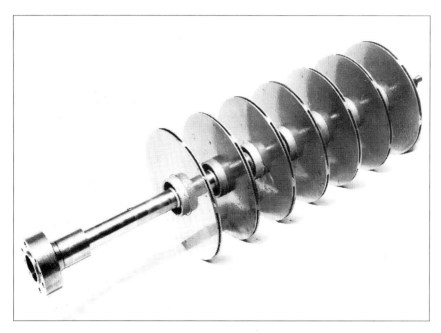

Figure 1:   Rotating assembly of a 7 stage unit

Figure 2:   A 2" x 2"x 16" single stage low flow, high head pump produced in 1971 for Shell, Stanlow

Figure 3:   'Milled for solid' casing, impeller and strainer

Figure 4:   'Milled from solid' open impeller

Figure 5: Tangential discharge casing 100 x 80-500 C.L. pump in UNS NO8825 material for produced water service

Mobil

AMEC

80x40 - 400 CL pump in "milled from solid" construction

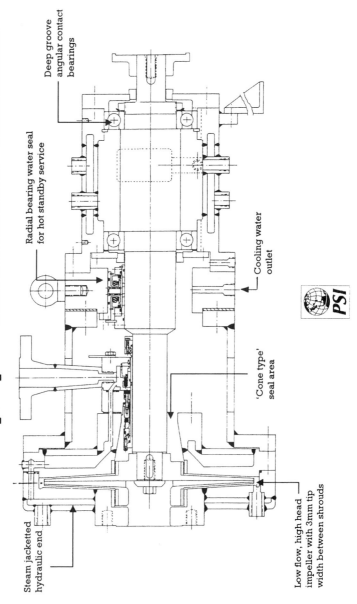

Deep groove angular contact bearings

Radial bearing water seal for hot standby service

Cooling water outlet

'Cone type' seal area

Steam jacketted hydraulic end

Low flow, high head impeller with 3mm tip width between shrouds

PSI

Figure 6: 80 x 40-400 C.L. pump for BP Coryton

Figure 7: Shrouded Impeller of 395 mm dia. with 3 mm tip width between shrouds of 80 x 40-400 C.L. pump

Figure 8: Shrouded impeller of 395 mm dia. with 3 mm tip width between shrouds of 80 x 40-400 C.L. pump

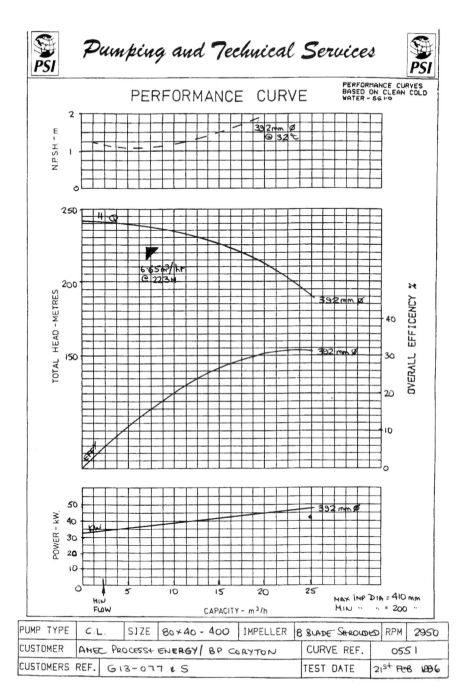

Figure 9: Performance curve of 80 x 40-400 C.L. pump for BP Coryton

Figure 10: One of pair 80 x 40-400 C.L. pumps for BP Coryton

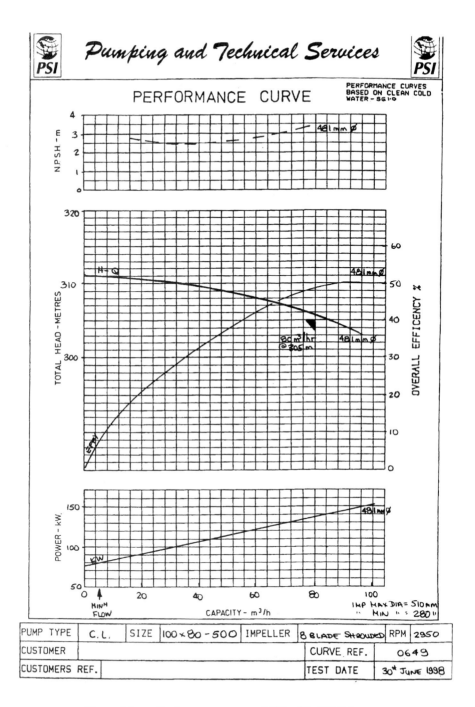

Figure 11:   Performance curve of 100 x 80 -500 C.L. pump

Mobil

65x40 - 250 CL pump with open impeller

AMEC

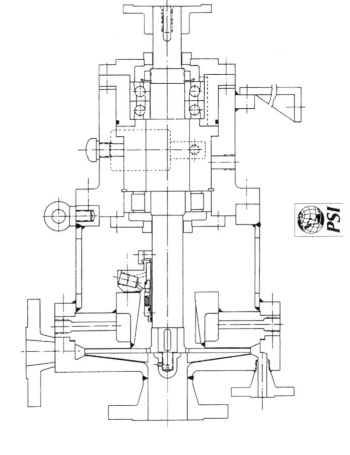

Figure 12: 65 x 40 250 C.L. pump with open impeller showing all wrought construction

Figure 13: 65 x 40-250 casing foot mounted, all wrought construction pump, magnetic drive, for BP Grangemouth

C556/022/99

Responsiveness leadership in compressor and turbine manufacturing

I M ARBON
Peter Brotherhood Limited, Peterborough, UK

SYNOPSIS

Although the development of the latest 'widget' is the normal topic of papers at a congress such as this, I believe that issues of quality assurance, operational reliability, machine availability factor and, most of all, credibility of reduced manufacturing lead-times are the crucial issues facing users of fluid machinery.

This paper examines my own company's success in achieving significant lead-time reductions through a process we call 'responsiveness leadership'.

The paper also examines several of the major technical management issues facing the fluid machinery industry - issues such as assessing the suitability of API Standards, complying with the European Directives, deciding whether or not to apply the CE-mark, and evaluating the impact of CRINE. In addition, the paper addresses more general business management issues such as the usefulness of generic strategies, the advantages of supply-chain partnering, implementation of project team-working and the overall management of change.

All of these will be illustrated by practical examples from my own company, which has seen a huge organisational transformation since mid-1996. It has changed from being merely a 'supplier of product' to become a 'provider of solutions' to its customers.

1   CURRENT STANDARDS IN THE FLUID MACHINERY INDUSTRY

Throughout the 1990s, as oil and gas companies and power generators around the world have been forced to increase their international competitiveness, there has been pressure to significantly reduce the level of investment in capital equipment for their production plants.

In particular, these major users have driven down the prices of fluid machinery such as compressors, turbines and pumps.

In fact, the industry has now been a 'buyer's market' for almost two decades and cost reductions are increasingly being imposed on manufacturers by their customers. Manufacturers have been forced to work with profit margins significantly lower than those achieved in the 1970s. As a direct result, a number of manufacturing companies have had to amalgamate in order to survive or have been forced out of the business altogether.

There has, understandably, been little enthusiasm on the part of these manufacturers to further erode their margins, especially when the user companies are predominantly found among the Fortune 100 largest companies in the world. However, many recent developments in the industry are actually designed to increase, rather than reduce, manufacturers' costs.

1.1   Suitability of the API Standards

Everyone here will be familiar with the various American Petroleum Institute (API) Standards, especially the API 61x series which so dominates our industry.

During the 1980s, I had the opportunity of working with the sub-committees dealing with API Standards 614[1], 617[2], 618[3] and 619[4]. The main intention of these standards is to provide the user of the machine with a safe and reliable piece of equipment and to give long periods of service between overhauls. These are admirable objectives which, sadly, will not be achieved through this means.

The way in which the standards were developed is interesting. The sub-committee meetings were divided into three categories of representative - users, contractors and manufacturers. Practically all of the decisions were made by user representatives based on largely anecdotal evidence of problems they had experienced over the (usually lengthy) span of their careers.

When a change to a standard was proposed and put to the vote, only user and contractor representatives were allowed to vote; manufacturer representatives were not! An example of the problems that this system could cause is that - until the publication of the third edition in 1997 - screw compressors complying with API 619[5] were required to have separate bearing housings, even though not a single manufacturer in the world had such a screw compressor design!

Another frequently encountered problem with API Standards is that they are differently applied in various parts of the world.

In the USA, the home of these standards, they are generally interpreted as 'recommendations' whereas in Europe, where industry is much more regulatory-minded, they are often applied as rigid 'standards'. This explains the apparent anomaly that European manufacturers usually list fewer 'exceptions to API' in compressor specifications than do their American counterparts. This is despite the fact that API Standards are unashamedly American and were originally devised as a 'barrier to trade'. The result of the different approaches on the two sides of the Atlantic has been that compressor and turbine design innovation has been encouraged in the industry in North America but discouraged in the same industry in Europe.

Despite these standards having always been the user's imposition on the manufacturer, there has been a steady dilution of the wording used as lawyers have begun to play an increasingly influential role. Much of the time at sub-committee meetings nowadays is spent in ensuring that no real 'specifications' are made. This is to ensure manufacturers gain no legal claim against API in the litigious USA, where it is possible for a manufacturer with a faulty product to sue API over alleged deficiencies in the standards that have led to the product defect!

Not only is 'specifying' language avoided, even terms such as 'recommendation' are not favoured. For example, in one recent meeting (1998) the proposed term "should be" was replaced by the rather meaningless "may be" in one clause.

Summary
Although API Standards have been well-known and highly-regarded in the industry for many years, I believe that they are often too rigidly applied in inappropriate situations. The problems which may be encountered in a European refinery in the future are unlikely to be the same as those encountered in a North American refinery in the past, although it is the latter which form the basis of the API Standards.

The API Standards are no substitute for sound technical judgment on the part of the user's and/or contractor's engineering staff. Furthermore, particularly in the European market, their avoidance of specifying language will ultimately lead to much confusion. In terms of product cost, there could be a huge difference between one supplier which holds rigidly to the terms of the standards and another supplier which interprets the more costly requirements simply as recommendations.

1.2   Compliance with European Directives
European Directives are developed by the European Commission. The directives of particular interest to the fluid machinery industry are ostensibly intended to enable the free transfer of goods throughout the European Union (EU). Their purpose is to eliminate technical barriers to trade by harmonising the national laws of member states regarding the design and manufacture of machinery and the assessment of the conformity of machinery with the directives. The directives are passed into law by the individual member states of the EU and are binding across national boundaries.

The directives appear to be a major step in the right direction because one of the main barriers to trade within Europe in the past has always been the insistence - by certain member states - on maintaining their national codes. However, to conform with the directives, equipment supplied must:

- Be safe;
- Meet essential safety requirements covering design, manufacture and testing;
- Satisfy appropriate conformity assessment procedures;
- Carry the CE-mark and other information.

The directives which are currently legally binding for equipment such as compressors, turbines and pumps supplied in the fluid machinery business are:

a) The Machinery Directive (89/392/EEC)[6], as amended by 91/368/EEC. In the UK, this was converted into legislation as The Supply of Machinery (Safety) Regulations 1992[7], Statutory Instrument (S.I.) 1992 No. 3073 which came into force on 1 January 1993. This directive was further amended by 93/44/EEC and 93/68/EEC, which were converted into UK Health and Safety Regulations, as S.I. 1994 No. 2063.

b) The Low Voltage Directive (73/23/EEC)[8] as amended by 93/68/EEC.

c) The Electro Magnetic Compatibility (EMC) Directive (89/336/EEC)[9] as amended by 92/31/EEC and 93/68/EEC.

d) The Pressure Equipment Directive, which was adopted by the European Parliament and the Council on 29 May 1997 as Directive (97/23/EC)[10]. As far as the UK is concerned, compliance with the directive is optional until 28 May 2002 (although national standards must still be adhered to); thereafter compliance with the directive is **mandatory**.

Although, obviously, no-one could disagree with legislation that leads to the increased safety of equipment supplied for use in the fluid machinery industry, there must be a clearer recognition within the industry of the additional cost burden that this places on manufacturers.

For example, there is a requirement, under each of the directives, for the manufacturer to take responsibility for a 'hazard analysis' for the equipment sold, to ensure its compliance with the health and safety legislation applicable in the country of use. Previously, in this industry, it had normally been the responsibility of the user to perform a hazard analysis (or a 'hazop' study) for the complete plant being installed. Although most manufacturers, as the designers of the equipment, are in the best position to perform such an analysis, it has to be recognised that there are both time and cost implications arising from this new legislation.

Summary
Although European Directives are still regarded as 'optional' by some manufacturers, and even by some users and contractors, they are now binding legislation in most EU countries. Whether we like them or not is immaterial - we have them, they are here to stay, and they are still growing in number.

Compliance with the Machinery Directive is already fairly onerous for manufacturers; compliance with the Pressure Equipment Directive will be much more so. Although compliance is a legal requirement, there is no doubt that compliance has a significant cost, which ultimately has to be passed on to the customer.

1.3   Whether or not to apply the CE-mark
The first three European Directives noted above have already passed into legislation in most EU countries. This means that for a supplier to sell a product to a customer within the EU, conformity with the directives is a legal requirement:

"Products meeting the requirements must bear CE marking, which means that they can be sold anywhere in the 15 member states of the European Community ...as well as Norway, Iceland and Liechtenstein." [11]

As already seen, the directives are primarily concerned with the safety of the operator of the equipment supplied, which seems to be a reasonable requirement. It is surprising, therefore, that many manufacturers of fluid machinery have decided that they do not need to apply the CE mark to the equipment they have manufactured. The reason usually given for this is that a compressor, for example, is not a machine until it is being driven by a motor or other driver!

Summary

This decision by some manufacturers not to apply the CE mark appears to be absurd. It must inevitably lead to suspicion that such a manufacturer is attempting to gain competitive advantage by not incurring the time and cost penalties of complying with the legislation. Although there is no attempt here to identify manufacturers which take this position there is a clear need for 'caveat emptor' among users and contractors preparing comparative bid evaluations.

1.4   Impact (or not) of CRINE

The Cost Reduction Initiative for the New Era (CRINE) has, in a sense, been brought about by the rigid adherence over many years to the API Standards mentioned earlier, as well as the more recent effect of the legislative requirements of European Directives. Again, in principle, the objective of trying to reduce the cost of equipment supplied for use in the fluid machinery business is a laudable one, which should benefit every company. However, CRINE has not been the enormous success that had been expected and the reasons for this need to be examined.

Let us examine, as an example pertinent to this industry, the CRINE Functional Specification for a "Gas Compression Package"[12]. Although the narrative section of this document runs to a mere four pages, there are several aspects of the specification which seem to be totally at variance with the fundamental concept of cost reduction - the 'CR' of CRINE:

a) The opening line of the specification states that it "defines the functional requirements for supply of a gas compression package for installation and operation on an offshore facility...". Not until the various appendices to the specification are reached does it become evident that the only type of offshore gas compressor recognised by CRINE is a centrifugal one! Given the large number of rotary screw and reciprocating gas compressors operating on offshore facilities, even in the North Sea, this is an astonishing oversight. Obviously, for certain duties, a centrifugal compressor is necessary, but if it is not necessary then it is an extremely expensive method of achieving gas compression - ie, it is cost-increasing and not cost-reducing.

b) The appendices consist, almost entirely, of copies of API data sheets. Since everyone involved in this industry is extremely familiar with API data sheets and knows exactly where to find them in the various API Standards to which they refer, they do not need to be reproduced in another document. Also, as already mentioned, because API is very conservative in its approach it tends to stultify innovation and, most importantly, it will do absolutely nothing to reduce the cost of the products.

c) Echoing the requirements of the European Directives, in paragraph 3.2 the specification states that "the supplier is responsible for ensuring that all equipment and components provided are suitable for the operating conditions stated, and for the listed fuels and utilities." The obvious problem here is that the supplier has no control whatsoever over the user's 'operating conditions'; indeed, everyone here will know of projects where even the user had little knowledge of, or control over, the operating conditions of its own site! For the supplier to honestly and truthfully assume responsibility for the suitability of the equipment offered would be a very time-consuming, and therefore costly, task. This cost would have to be passed on to the user of the equipment.

d) Probably the greatest concern that manufacturers should have concerning this CRINE document is in paragraph 5.1, "Documents Required with Bid" (see Fig. 1).

Fig. 1 Documents <u>required</u> with Bid

- Completed data sheets, utility requirements, weight details and other interface data.
- Drawings detailing package envelope, including installation / maintenance requirements.
- Technical Specification.
- Quality Plan.
- Manufacturing and Delivery Schedule.
- Reliability / Availability Data.
- Comments / Deviations / Exceptions.
- Commissioning Spares.
- Recommended Spares List (Operating).
- Overall scope definition (i.e. completed Appendix A).
- Identification of major hazards.

Fig. 1 Documents <u>required</u> with Bid (CRINE Initiative)

The list includes not only everything the average manufacturer already must include in a bid for equipment for the fluid machinery business but much more besides!

A quick estimate of the time involved in preparing this level of information for a typical bespoke gas compressor package suggests about 200 design hours for a single bid. Most customers still prefer to obtain at least three competitive offers for any project. This means a manufacturer is expected to undertake this level of engineering work, at its own expense, before it has an order to commence work and with only a one-in-three chance of ever winning the order! Since most manufacturing companies in this industry work on relatively small margins, it is evident that the cost of this redundant work has to be passed on to the customer. This is a further example of a huge cost-increase resulting directly from a document which sets out to promote cost reduction.

Summary
Although based on laudable premises, there is little evidence that the various CRINE publications do anything for the stated aim of cost reduction. Although ostensibly written to permit the use of manufacturer's standard offers, the net result of these additional specifications is that the machine offered must be 'API plus'.

1.5   Conclusion
From the foregoing analysis it is evident that the three main series of specifications designed to protect the user in the fluid machinery industry - namely, API Standards, European Directives, and CRINE - will all add hugely to the manufactured cost of the product.

The remainder of this paper is devoted to looking at ways in which a manufacturing company entering the new millennium can still develop innovative strategies which enable it to offer cost reductions to its customers.

2   DEVELOPMENT OF COST REDUCTION STRATEGIES

2.1   Usefulness of generic strategies

Over the past two decades it has become fashionable in business schools and consultancies to develop so-called 'generic strategies' for easy use and rapid implementation by industry. Usually these 'generic strategies' are extremely simplistic and take no account of the 'environmental scanning' which is indispensable before the implementation of an effective strategy by a company. de Kare-Silver (1997)[13] seems to be the first writer to expose the very serious limitations of the various generic strategy models, from the Boston Matrix[14] of the early 1970s through the concepts of Porter's Generic Strategy (1980)[15], PIMS (1987)[16], 'core competence' (1990)[17], even up to the 1994 concept of 'Parenting'[18].

This view is well summarised by Harari (1994)[19]:

> "Generic strategies, so popular over the past decade, are not really so; [they are] worked out by academics as after-the-fact case studies!"

In total contrast to the 'generic strategy' concept, De Wit & Meyer (1994)[20], in their case study of the Kao Corporation, quote Dr Maruta, its CEO, as saying:

> "Yesterday's success formula is often today's obsolete dogma.   We must continuously challenge the past so that we can renew ourselves each day"

For example, let's analyse the most commonly applied of the above models, namely Michael Porter's (a Harvard academic!) 'Generic Strategies Diagram', see Fig. 2.

Fig. 2 Porter's "Generic Strategies"

Fig 2. Michael Porter (1985) Competitive Advantage

This model is slavishly followed by many business schools but does not seem to be based in industrial reality - certainly not in the fluid machinery business. Porter's book Competitive

Advantage[21], from which the diagram was taken, was published back in 1985. Porter views 'cost leadership' and 'differentiation' as mutually exclusive strategies – "a firm that tries to pursue both is 'stuck in the middle'". Given that by 'cost leadership' he is referring to the 'cost of production', it is most astonishing that he makes no mention whatsoever of the impact of currency exchange rates on 'cost leadership'. Yet, to any British manufacturing exporter, in any industry sector in 1997/8, this was far and away the most important criterion for consideration.

In an eighteen month period during which the exchange rate was close to 3.00DM/£ it became extremely difficult for British manufacturing companies to compete, on a 'cost leadership' basis, even with the European manufacturing giants. Most UK manufacturing companies believed that they needed a maximum rate of about 2.55DM/£ to be cost competitive[22].

Nor does Porter consider the hugely varying cost of labour internationally. Fig. 3 shows the UN/USDoL's comparison of 1995 and 1997 figures of the relative average cost per hour of production workers (USA = 100)[23].

Fig. 3 Comparative Production Costs

Country	1995	1997
Germany	152	126
Japan	139	106
France	112	96
USA	100	100
UK	81	86
Korea	39	37
Taiwan	32	30
Mexico	17	19
China	4	5

Fig. 3 Average cost per hour of production workers

It is clear from Fig.3 that a company can achieve significant 'cost leadership' advantage simply by moving its point of manufacture, although by Porter's definitions this would seem to be more akin to 'product differentiation'. This is despite the fact that Porter states that 'cost leadership' and 'product differentiation' are mutually exclusive strategies!

2.2   Adopting a strategy for 'responsiveness'
Given the lack of applicability of the 'generic strategies' to the capital equipment business, one interesting discovery is the differentiation diagram, reproduced in Fig. 4, showing 'technology leadership', 'cost leadership' and 'responsiveness leadership'. This shows quite clearly that, in the capital equipment business, cost leadership is a form of product differentiation, not an alternative to it. This diagram has been used as the basis of the new strategy in my company.

    C556/022 © IMechE 1999

The purpose of this diagram is to show the three main categories used to differentiate a product in the capital equipment business. The question being answered by this diagram is: "Why would a customer buy one manufacturer's product rather than another?"

The reasons are:

- Technology leadership, i.e., higher efficiency, more advanced materials, improved reliability of operation, etc.
- Cost leadership, i.e., the lowest capital cost, although probably not the lowest operating cost.
- Responsiveness leadership, i.e., the quickest response to the customer's needs both before the order is placed and during the period of order fulfilment.

The diagram shows these three characteristics $120°$ apart. Experience has shown that, in this industry, no company can focus on all three characteristics at the same time. For example, a company which has invested heavily in R&D is unlikely to achieve 'cost leadership' and will probably struggle to achieve 'responsiveness leadership' The optimum result ensues from focussing on <u>one</u> of the characteristics, although two may be achieved at the same time if some degree of dilution of effect is acceptable.

Fig. 4 'Leadership' Diagram

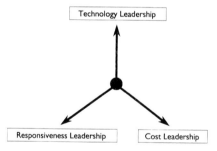

Fig. 4 "Leadership Diagram"   Source: Walker, A-C Compressor, 1995[24]

The purpose of this diagram is to show the three main categories used to differentiate a product in the capital equipment business. The question being answered by this diagram is: "Why would a customer buy one manufacturer's product rather than another?"

The reasons are:

- Technology leadership, i.e., higher efficiency, more advanced materials, improved reliability of operation, etc.

- Cost leadership, i.e., the lowest capital cost, although probably not the lowest operating cost.
- Responsiveness leadership, i.e., the quickest response to the customer's needs both before the order is placed and during the period of order fulfilment.

The diagram shows these three characteristics 120° apart. Experience has shown that, in this industry, no company can focus on all three characteristics at the same time. For example, a company which has invested heavily in R&D is unlikely to achieve 'cost leadership' and will probably struggle to achieve 'responsiveness leadership' The optimum result ensues from focussing on <u>one</u> of the characteristics, although two may be achieved at the same time if some degree of dilution of effect is acceptable.

Since, in a worldwide context, my own company is relatively small, we decided not to focus on technology leadership but to concentrate on being a 'fast follower' in this area. Similarly, with technically-acceptable compressors and turbines now being manufactured in countries like Brazil and India, it is futile to adopt a cost leadership strategy. Consequently, we have adopted responsiveness leadership as our primary strategy although we have found that this has also resulted in some very significant cost savings.

2.3   Responsiveness leadership
The key issue for us was finding a way to beat the competition. Typically, steam turbine and/or gas compressor projects have overall lead-times of 55 to 60 weeks. Our strategy was to gain a competitive advantage by reducing this period by at least ten weeks - and ultimately reducing it by a further ten weeks to between 35 and 40 weeks.

The traditional way to handle contracts in this industry sector is to appoint a contract engineer responsible for ensuring that all engineering work is carried out, to enable the factory to produce the complete machine in the allotted lead-time. Normally the first phase engineering work is carried out over a twelve to sixteen week period between the date of order and the vendor co-ordination meeting (VCM) where the contract-specific drawings are reviewed with the customer and the precise scope of supply agreed.

The VCM has been described as "the time when we see if there is any similarity between what we think we've sold and what you think you've bought"! Until this point is reached and the scope of supply is clearly defined, it is impossible to commit to making the long lead-time capital purchases required for the job. Typical lead-times for large electric motors or generators are between 35 and 40 weeks which is usually the determining factor (i.e., the critical path item) for contracts requiring such equipment.

Clearly, if the time between date of order and VCM can be eliminated or significantly reduced, the overall target of lead-time reduction can be readily achieved.

To do this we have adopted a policy of completing most of the first-phase engineering work before the purchase order is received. By this means, the contract-specific drawings can be completed in the first two weeks of the contract and the VCM held three to four weeks into the contract.

Our experience so far has shown that it is possible to achieve a fairly well-defined scope of supply at the time of order placement. Also, supply-chain partnerships established with major manufacturers of electrical machines mean that these companies have also completed their first-phase engineering work by the time of the VCM, enabling the order for the motor or generator to be placed immediately afterwards. This early completion of the engineering work has also enabled the electrical machine manufacturers to reduce their lead times, which further helps us in achieving our target reductions.

During 1997-8, our company was restructured with a focus on customers/markets into five strategic business units (SBUs); each SBU now has full responsibility for its own profit & loss statement and sales and marketing activities. The new organisation of the company is shown diagrammatically in Fig. 5.

Fig. 5   Peter Brotherhood Ltd
SBU Web Structure

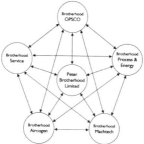

Fig. 5 "Five SBU Diagram"   Source: Arbon, Peter Brotherhood Ltd, 1997[25]

The SBU which handles our gas compressor and steam turbine products is called 'Process & Energy'; this SBU has since been re-structured from a traditional hierarchical organisation into what we call the 'three-group' concept, see Fig. 6.

Before this, much development work was done on a specific contract which was costly in terms of time. Now, all development, standardisation and rationalisation of the product is within the 'Strategic Technology Group', see Fig. 6. This enables the 'Order Fulfilment Group' to focus on executing the contract in accordance with the customer's requirements and in the most cost-effective way possible. At the same time, the engineering capabilities of the 'Order Winning Group' have been strengthened so that project-specific general arrangement drawings (GAs) and pipework and instrumentation diagrams (P&Is) accompany any final-stage proposal, thus clearly defining the scope of supply even at tender stage.

Fig. 6   The "Three Group" Concept
in the Process & Energy SBU

Fig. 6 "Three Group Diagram"   Source: Arbon, Peter Brotherhood Ltd, 1997[25]

Several changes in the factory (now another SBU called 'OPSCO' - see Fig. 5) have resulted from this new strategy. The most important is that the quasi-JIT philosophy formerly used in the company has been abandoned .   Although often touted as an improvement in manufacturing strategy, this can often be an accountant's naive method of neutralising cash flow. Every component required for the contract is now ordered as early as possible and held in stores until all the parts required for building the package are available. Only then does the compressor or turbine package start its build program. The 'dwell' time on the shop floor is therefore extremely short and time is not booked to the contract unnecessarily.

2.4   Advantages of supply-chain partnering

The concept of 'supply-chain partnering' has grown out of the booming microprocessor and electronics sector. It has been adopted by certain other industrial sectors but not, to any great extent so far, by the fluid machinery sector.

The only evidence of 'partnering' in this industrial sector of which I am aware has resulted from the CRINE initiative discussed above.   Some of these 'partnerships' have run into difficulties because of a fundamental misunderstanding of the concept. For example, a user or contractor may develop a supply-chain partnership with a compressor supplier with extensive experience in ammonia plants. The next large contract is perhaps for an ethylene plant, where the compressor supplier has little experience, so the purchaser has to rely on competitive tendering and the previous partnership comes to an end. Clearly, then, such 'partnerships' need to be much more flexible in an industry obsessed with 'experience'.

Supply chain partnerships (SCPs) are more readily established as far as the fluid machinery manufacturer is concerned. As outlined in 2.3 above, my own company has been quite successful in establishing SCPs with suppliers of components such as electric motors/generators, heat exchangers, oil consoles and the like. SCPs of this type enable significant levels of standardisation and consequent reduction of lead-times.

2.5   Implementation of project team-working

Prior to the original formation of the SBUs in August 1996, our company operated a traditional hierarchical structure, with the Compressor Division completely separate from the Turbine Division. Each had its own design engineering, electrical, tendering and contracts departments; the drawing office and sales departments were separate divisions and the marketing function did not exist! There was little effective interface between the departments.

All of the above functions were drawn together as "Process & Energy" in the SBU restructuring (Fig. 5) and, for the first time in the company's history, all the people in these functions sat at adjacent desks. The next stage was to move the people in the SBU into three very focussed, yet intersecting workgroups, as was shown in Fig. 6. These are:

- 'Order Winning' which includes sales, tendering and applications engineering.
- 'Order Fulfilment' which includes contract management, design engineering, electrical and CAD.
- 'Strategic Technology' which is a completely new concept that combines elements of all of the foregoing functions plus marketing.

The remit for 'Order Winning' is to focus on winning new business for the SBU with a minimum level of estimated profit margin. 'Order Fulfilment' concentrates on executing the contract to the customers' complete satisfaction, while maintaining or growing estimated profit margins. 'Strategic Technology' supports the other two teams by focussing on 'cost leadership' (by value engineering, outsourcing, improved manufacturing processes, etc) as well as new product and market development. The three teams were formed on 1 May 1997 and have operated very successfully since.

The most radical development here was the Strategic Technology group. This was not simply an updated R&D department but a genuine attempt to have the company develop products which the customers want to buy, rather than those which the company wants to make. This was the reason for including the marketing function within this team, rather than in Order Winning, which would be more conventional.

3   SUMMARY AND CONCLUSIONS

3.1   Topics

This paper has critically examined several of the more challenging technical issues encountered in the fluid machinery industry - issues such as the suitability of API Standards, compliance with European Directives, whether or not to apply the CE-mark, and the impact (or lack of impact) of CRINE. Although all of these standards and specifications have important roles to play in this industry, it has to be recognised that strict adherence to such codes will almost certainly have a huge cost impact without necessarily improving the safety and operational reliability of the machine. At the same time as insisting on rigid compliance, the industry is still trying to find ways to achieve significant cost reduction.

3.2   Solutions

The paper has then proceeded to demonstrate how implementation of some of the business and management issues facing today's manufacturers of fluid machinery can be utilised to provide the required cost improvements, particularly through lead-time reduction. Some of the concepts prevalent in other industrial sectors, the usefulness of generic strategies, advantages of supply-chain partnering, and the implementation of project team-working, are also reviewed with a view to their use in the fluid machinery business.

3.3   Conclusion

The fluid machinery market is looking for reliable products, safe in operation, providing long intervals between overhauls and with extremely high availability factors; yet all of this is required at the lowest possible price and with the shortest possible delivery lead-times. This paper has demonstrated that the major industry standards and procedures, such as API Standards, European Directives and even CRINE, while undoubtedly contributing to the former objectives, are actually counter-productive to the achievement of the latter objectives of lower prices and shorter deliveries.

The obvious conclusion to be drawn is to develop the concept of 'partnering' between Supplier and Customer, introduced to a degree in CRINE, to the point where it can achieve all the above objectives. This would involve both the Supplier and Customer in the partnered development of a 'fit-for-purpose' machine providing the cost and lead-time savings the Supplier can offer while still achieving the safety and reliability the Customer needs. The Supplier's willingness to apply the CE-mark to its product is clearly central to this proposal; his taking into account any operating hazards in the machine is a far greater safeguard than the rigid adherence to standards and specifications developed by others.

REFERENCES

1   American Petroleum Institute Standard 614 (1984)
     Lubrication, Shaft Sealing & Control Oil Systems, (2nd ed), API, Washington, DC, USA

2   American Petroleum Institute Standard 617 (1988)
     Centrifugal Compressors, (5th ed), API, Washington, DC, USA

3   American Petroleum Institute Standard 618 (1986)
     Reciprocating Compressors, (3rd ed), API, Washington, DC, USA

4   American Petroleum Institute Standard 619 (1985)
     Rotary-Type Positive Displacement Compressors, (2nd ed), API, Washington, DC, USA

5   American Petroleum Institute Standard 619 (1997)
     Rotary-Type Positive Displacement Compressors, (3rd ed), API, Washington, DC, USA

6   Official Journal of the European Communities (1989)
     Directive 89/392/EEC of the European Parliament and Council, EURop, L-2985 Luxembourg

7   Statutory Instrument No. 3073 (1992)
     The Supply of Machinery (Safety) Regulations 1992, HMSO, London, UK

8 Official Journal of the European Communities (1973)
   Directive 73/23/EEC of the European Parliament and Council, EURop, L-2985 Luxembourg

9 Official Journal of the European Communities (1989)
   Directive 89/336/EEC of the European Parliament and Council, EURop, L-2985 Luxembourg

10 Official Journal of the European Communities (1989)
   Directive 97/23/EC of the European Parliament and Council, EURop, L-2985 Luxembourg

11 Department of Trade and Industry (1998)
   Guidance Notes on the EC Pressure Equipment Directive (97/23/EC), London, UK

12 CRINE Network (September 1994)
   Functional Specification - Gas Compression Package, CRINE Ltd, London, UK

13 de Kare-Silver, Michael. (1997), Strategy in Crisis, (1st Ed.)
   London, UK: MacMillan Press

14 Henderson, Bruce. (1984), The Logic of Business Strategy
   Harper & Row, USA: Boston Consulting Group

15 Porter, Michael. (1980), Competitive Strategy (1st Ed.)
   New York, NY, USA.: Free Press

16 Buzzell, R. D. & Gale, B. T. (1987), The PIMS Principles, (1st Ed.)
   New York, NY, USA.: Free Press

17 Prahalad, C. K. et al (May/June 1990), The Core Competence of the Organisation
   Boston, MA, USA: Harvard Business Review

18 Goold, K., Campbell et al (1994), Corporate-Level Strategy, (1st Ed.)
   London, UK: John Wiley

19 Harari, Oren. (August 1994), The Hypnotic Danger of Competitive Analysis
   Management Review, USA

20 de Wit, Bob & Meyer, Ron. (1994)
   Strategy, Process, Context (1st Ed.), St Paul, MN, USA: West Publishing Company

21 Porter, Michael E. (1985): Competitive Advantage; Creating and Sustaining Superior Performance, New York, NY, USA: Free Press

22 Engineering Employers' Federation (1998)
   Business Trends Survey and Economic Review 1998, London, UK

23 United Nations (1995 & 1997)
   Statistics, New York, NY, USA: UN Statistical Office;
   U. S. Department of Labor (1995 & 1997)
   Bureau of Labor Statistics, Washington, DC, USA

24 Walker, Gary. (1995), Responsiveness Leadership
   Appleton, WI, USA: A-C Compressor Corp

25 Arbon, Ian M. (1997)
   Company Conference Presentation, Peterborough, UK: Peter Brotherhood Ltd,.

C556/001/99

Rotordynamic design considerations for a 23 MW compressor with magnetic bearings

A B M NIJHUIS
Delaval Stork, Vof, The Netherlands
J SCHMIED
Delta JS AG, Switzerland
R R SCHULTZ
Glacier Magnetic Bearings, USA

This paper describes a very demanding magnetic bearing application. A multi-stage centrifugal compressor, with a maximum discharge pressure of 130 barg, driven by a variable speed synchronous electric motor, with an operating speed range from 600 rpm to 6300 rpm. The requirements for the rotordynamic performance of the compressor are defined. Special engineering tools, covering the mechanical aspects as well as the electronic controller for the magnetic bearing system, are needed to comply with the requirements. The results of the calculations and an experimental verification are given.

1   INTRODUCTION

The industry, through it's increasing demand for advanced, versatile but robust rotating machinery, requires the application of modern, environmentally friendly technologies. Active magnetic bearings are gaining more and more acceptance and are becoming a serious feature for the industry to consider. A very prominent example is a compression unit that features a centrifugal compressor, driven by a 23 MW variable speed synchronous electric motor, both using active magnetic bearings. Since dry gas seals are used in the compressor, it follows that this unit is completely dry, therefore no lubrication or seal oil is needed. The paper covers the rotor dynamic aspects of the compressor and is focused on the analytical methods applied during the design phase. An experimental verification of the rotor response measured during extensive shop testing is included.

2   DESCRIPTION OF THE NATURAL GAS PLANT AND THE COMPRESSOR

The Groningen natural gas field, which is located in the northern part of The Netherlands, is operated by NAM B.V., a shareholder company of Shell Nederland B.V. and Esso Holding Company Holland Inc. The field has 29 production installations (clusters) spread over the province of Groningen (see Figure 1).

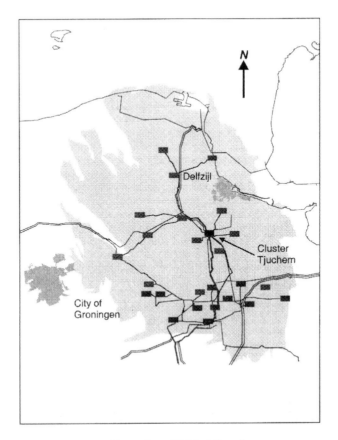

Fig. 1 Natural gas Field of Groningen

The Groningen Field is used as a peak flow supplier with continuous production swings, mainly due to national daily and seasonal demands. (see Figure 2).

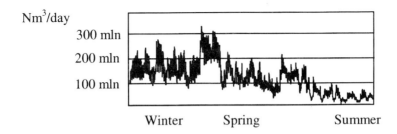

Fig. 2   Gasproduction over 6 Months in Groningen

Contractual commitments to deliver sufficient capacity at a supply pressure of 65 barg to the grid, require that compression units be installed in the near future. Feasibility studies concluded that compression upstream of the gas conditioning plant on all clusters would be the most economical solution. See Figure 3 for a schematic overview of the gas treatment plant.

The project is a good application for active magnetic bearings, since they facilitate extension of the operating speed range of the unit all the way down to 10%, and also allow for remote monitoring. The operating principle of magnetic bearings has been explained in various technical papers. The present limitations are also documented [1].

A cross section of the compressor is shown in Figure 4. The initial compressor is a 5-stage configuration. In the future the compressor will be revamped to an 8-stage configuration.

Tandem dry gas seals are installed at the shaft ends. Active magnetic journal bearings are applied in combination with self lubricated, auxiliary bearings. These auxiliary bearings are located at carefully chosen locations on the rotor, to prevent damage to rotor and stator parts during emergency rundown situations.

An active magnetic, double acting axial bearing is installed at the Non-Drive-End (N.D.E.) of the compressor. The power is supplied through a dry diaphragm type coupling.

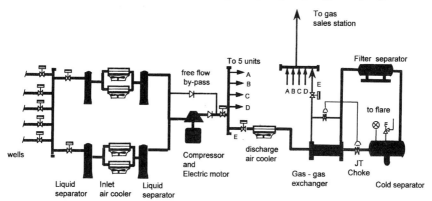

Fig. 3 Process Flow Scheme of Gas Treatment Plant, with Compression Unit

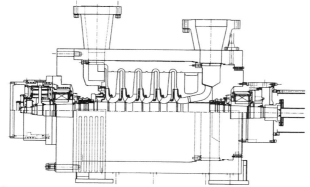

Fig. 4 Cross Section of the Compressor (Initial Configuration)

3   REQUIREMENTS FOR THE ROTORDYNAMIC PERFORMANCE.

The challenge for the design engineer for such an application is impressive and unique, since there is a lack of design consensus represented in the accepted industrial guidelines and standards. API 617 [2] deals only with fluid film bearings and contains no specific guidelines and acceptance criteria as reference data for both customer and supplier. Appendix J of this standard does contain generalised application considerations for active magnetic bearings. The application of active magnetic bearings mandates that the design engineer expands his horizon into the area of Mechatronics. The rotordynamics of a compressor rotor, with traditional fluid film bearings, is limited to the tuning of the interaction between the rotor and the oil film in the bearings. Active magnetic bearings however require that the rotor, the bearings and also the bearing control system are properly accounted for, i.e. rotor, sensors, controller, amplifiers and electromagnets.

Delaval Stork has built up significant experience in the application of dry-dry technology (dry gas seals and active magnetic bearings) over the years, on both beam type and overhung type centrifugal compressors [3, 4]. Based upon this in-house expertise the following considerations for dry-dry rotordynamic performance have been developed:

1. The allowable vibration level at each radial magnetic bearing shall not exceed twice the API 617 para 2.9.5.5 limits throughout the operating speed range. Active magnetic bearings have an inherent lower stiffness compared to that of traditional fluid film bearings. This means that vibration levels for a given unbalance will be higher at the bearing location. Assuming these higher vibration levels, the design engineer can better optimise the control loop algorithm which in this case is quite essential given the extremely large operating speed range. Maintaining the vibration limits specified in API 617 would mandate high stiffness from the active magnetic bearing control system with inherent high dynamic currents to the electromagnets. This introduces the risks of a sensitive control loop and saturation of the power amplifiers even during small displacements.

2. The amplification factor for all modes within the operating speed range shall preferably be less than 2.5. This means that on a case to case basis, prudent judgement should be given to this design criterion. Specifying an amplification factor of 2.5 (equals a damping ratio of 20%) as maximum, introduces constraints for the design engineer in the optimisation of the control loop.

3. The bearing control system shall provide adequate damping for all modes to allow stable operation for all the conditions specified. This mandates not only positive damping over a large frequency range but also exact knowledge of all excitation mechanisms, with labyrinth induced cross coupling being the most dominant [5]. Unlike fluid film bearings, active magnetic bearings are not as forgiving in overload situations.

4. All modes which can be excited during operation shall be clearly observable and controllable, i.e. the rotor shall yield sufficient amplitude at the various bearing sensor locations.

5. The bandwidth of the control loop should be as small as possible, i.e. a low frequency roll-off is preferred. This makes the bearing control system less sensitive for natural frequencies above the roll-off frequency. In this way, all rotors, each of which may exhibit slight differences in natural frequencies, due to manufacturing tolerances, are not limited to use in one machine. The rotor bundles are not assigned to a specific field cluster, but may travel from one cluster to another, as part of the maintenance concept adopted for the project. This approach eliminates the need for on-site tuning.

4   TOOLS FOR THE ROTORDYNAMIC ANALYSES

Software tools suited for the engineering of magnetic bearing applications must comprise a structural part describing the rotor as well as a mechatronic part describing the magnetic bearing, the controller and amplifiers, and the combined rotor bearing system. The following combination of programmes was used for the rotordynamic analyses. For the structural part the comprehensive machine dynamics programme MADYN was used, and for the mechatronic part MEDYN (mechatronic system dynamics) was used [6]. The latter are functions, which were developed using the mathematical software package MATLAB. The tasks of MADYN and MEDYN as well as the communication between the two programmes are summarised in Figure 5.

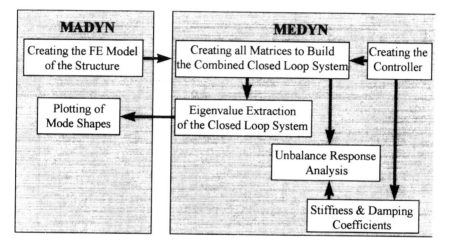

Fig. 5 Tasks of MADYN and MEDYN

In the structural part, the following effects are considered:
- Gyroscopic effects
- Rotor fluid interaction, such as labyrinth induced excitations
- Dynamics of flexible parts mounted on the rotor, such as disks

In addition to this the tools may also be used to account for the flexibility of stator parts, although this was unnecessary in this project, since all stator parts were sufficiently rigid.

The mechatronic part includes the modelling of:
- Non-collocation of sensors and actuators
- Negative stiffness of magnetic actuators
- Digital controllers
- Time delay for the digital controllers
- Characteristics of the hardware components of magnetic bearings, such as sensors and amplifiers
- Separate sensors as well as separate controllers for displacement and velocity

- Coupling of bearings by means of the controllers, such as separate controllers for the tilting and translation modes of rotors, which may be useful in the case of symmetric or almost symmetric rotors.

The mechatronic part also has a design tool for the controller, which allows combining standard controller components. The following components are available in analog as well as digital form:

- Base component, which is a modified PID controller with bandwidth limited phase lead cells
- First order filter
- Second order filter with a parallel proportional part
- Second order all pass filter
- General second order filter
- Notch filters
- Analogue Butterworth filters in case of digital controllers for anti-aliasing

5   BASIC ROTORDYNAMIC BEHAVIOUR

The rotor model used for all rotordynamic analyses is shown in Figure 6. The basic data of the rotor are as follows:

- Weight   2100 kg
- Total length   3845 mm
- Bearing span (between actuators)   2674 mm
- Maximum continuous speed   6300 rpm

The triangles in this figure indicate the positions of the bearing actuators. Each bearing has two sensors at each side of the actuator. Finally, the outboard sensors were used due to reasons that will be explained in section 6.

The disk at node 8 is the axial bearing disk. Its one nodal diameter vibration mode had to be considered in the rotor model, since its frequency of 490 Hz is in the range which must be considered for the magnetic bearing tuning (frequencies up to approx. 1000 Hz, also see section 6). To model this mode the disk is fixed to the rotor with a rotational spring. The stiffness of the spring is adjusted to yield the frequency of 490 Hz.

The investigation of the basic rotordynamic behaviour comprises the analysis of the natural modes for different bearing stiffnesses at different speeds. Their frequencies are shown in Figure 7. This diagram shows which modes are in the operating speed range, their separation margin from this speed range, and gives guidelines for the controller design, i.e. to show at which frequencies the controller must provide damping. Negative bearing stiffnesses are included, since the magnetic bearing can have such stiffness values at certain frequencies. In this application the first bending mode is in operating speed range, and requires special attention.

Figure 8 shows the first four bending mode shapes and the disk mode at zero speed. The long vertical lines indicate the actuator position and the short lines the sensor positions.

It can be seen that the disk mode and the rotor modes interact. In the second and higher bending modes the flexible disk tilts more than a rigid disk would. This has the effect of considerably increasing the gyroscopic effect. As a consequence the frequency difference between forward and backward whirl increases as well as the frequency regions where the bearing must provide damping.

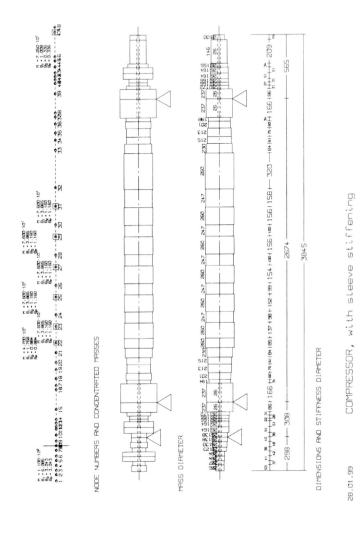

Fig.6 Rotor Model

COMPRESSOR, with sleeve stiffening

28.01.99

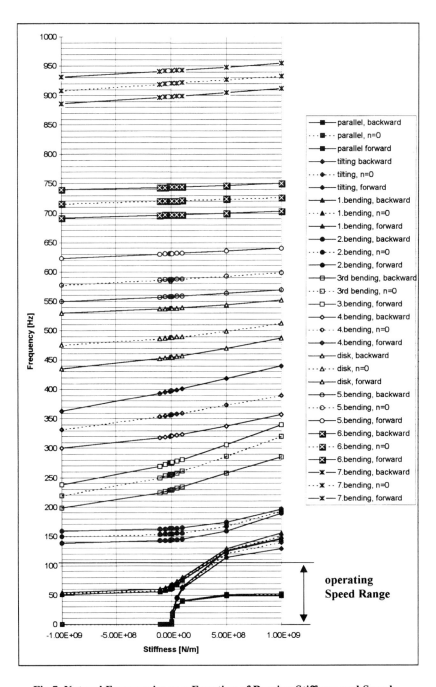

Fig.7 Natural Frequencies as a Function of Bearing Stiffness and Speed

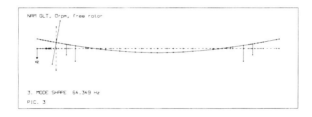

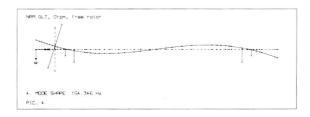

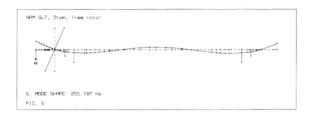

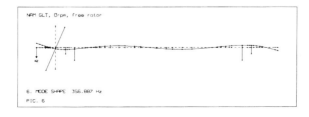

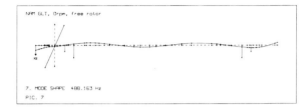

Fig.8: Natural Modes of the Free Rotor at Standstill

6   THE CONTROLLER DESIGN

The aim of the controller design is to provide good damping for all modes in the speed range of the machine and sufficient damping for higher modes in order to keep them stable. The bearing stiffness must also be high enough for the rotor to resist fluid forces, which normally have low frequencies. Additionally the controller must not create high currents (i.e. high forces) at high frequencies in order to prevent saturation of the power amplifiers due to noise (also see section 3).

Fig. 9 shows the magnetic bearing transfer function (sensor displacement to bearing force) including the sensor, the controller, the amplifier and the actuator. The actuator has a magnetic pull, which can be modelled as a spring with negative stiffness (k_s=1.57 10^7 N/m) and which has to be compensated for by the controller. The pull is not included in the transfer function as shown.

The bearing provides a damping force if the phase angle is in the following range:

$$0 \prec \varphi \prec 90^{\circ}$$

or

$$-180^{\circ} \succ \varphi \succ -270^{\circ}.$$

In the latter case the bearing stiffness is negative, which does not lead to instabilities at high frequencies. Damping is provided for frequencies below 125Hz and for frequencies above 270Hz. Between 125Hz and 270Hz a negative damping force is created. This is mainly due to a second order filter with its frequency at 160Hz, which has two functions:

1. To reduce the amplitude at high frequencies.
2. To drop the phase, which has a tendency to decrease with increasing frequency (due to the amplifier characteristic and the digital controller), below 180° into a positive damping region.

The filter was introduced in the frequency range of the second bending mode, because this mode has a node close to the outboard sensors. In fact, the node is even slightly inside the sensors (see figure 8). The negative damping force provides low rotor damping for this mode, which is sufficient. This is also the reason, why the outboard sensors were chosen as active sensors.

At around 18 Hz the phase is increased by an extra filter. This is to increase the damping of the rotor rigid body modes, which are in this frequency range and which are mostly affected by the destabilising labyrinth seal forces.

The overall controller transfer function is an eighteenth order polynomial that is designed using the bearing manufactures controller design software. To fully optimise the controller for this application the transfer function is synthesised with complex rather than simple poles and zeros. Digital controller hardware is essential to the implementation of this type of controller transfer function.

                                                    C556/001 © IMechE 1999

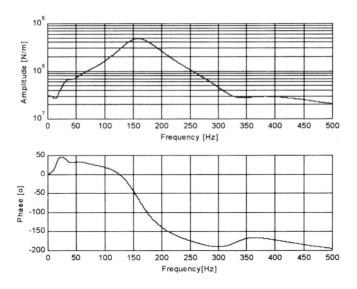

Fig.9 Magnetic Bearing Transfer Function (without magnetic pull)

7   ACHIEVED ROTORDYNAMIC PERFORMANCE

The eigenvalues (natural frequency and damping ratio) of the combined rotor bearing system at nominal speed and standstill are listed in Table 1. The table contains all eigenvalues which can be assigned to the rotor. Additionally some eigenvalues are caused by controller poles, which interact with the rotor. This interaction can change their frequency and damping ratio. The eigenvalues caused by the controller with a damping ratio below 20% are also shown in the table.

All modes below the maximum speed are very well damped. The first bending mode, which is within the operating speed range, has a damping ratio of 20%. This covers the API 617 requirement for compressors, although its application to compressors supported on magnetic bearings is controversial.

Fig.10 shows the calculated and measured bearing response of the DE bearing to an unbalance of the magnitude of G2, i.e. 3300 gramm-millimeters is applied at the thrust disk and 2700 gramm-millimeters at the coupling, at the same angular position. This distribution yields a good excitation for the first bending mode. The maximum force below the maximum speed of 6300 rpm remains below the bearing dynamic capacity of 20000 N peak-peak.

The agreement between measurement and calculation is good.

The calculated force at higher speeds (where the compressor is not required to operate) increases due to the controller dominated pole at 130 Hz, which has relatively low damping. Also the measured force remains at a high level up to maximum speed for the same reason.

Table 1: Natural Frequencies and Damping Ratio of the Rotor Bearing System

Mode	Frequency n=0	Damping Ratio n=0	Frequency n=6300rpm	Damping Ratio n=6300rpm
rotor parallel	12.6 Hz	25.8%	-12.6 Hz	25.8 %
			+12.6 Hz	25.8 %
rotor tilting	14 Hz	28.8%	- 13.8 Hz	28.8 %
			+ 14.1 Hz	28.9%
controller pole 35.8Hz, D=0.4	35.8 Hz	18.5%	- 35.7 Hz	18.3 %
			+ 35.9 Hz	18.7 %
rotor 1.bending	81.2 Hz	19.8 %	- 77.3 Hz	21.0 %
			+ 84.7 Hz	19.2 %
controller pole 161Hz, D=0.2	130 Hz	6.8 %	- 129 Hz	7.8 %
			+ 131 Hz	6.5 %
controller pole 161Hz, D=0.2	131 Hz	9.5 %	- 131 Hz	12.3 %
			+ 132 Hz	15.5 %
rotor 2.bending	155 Hz	1.4 %	- 143 Hz	1.5 %
			+ 164 Hz	1 %
rotor 3.bending	250 Hz	0 %	- 223 Hz	0 %
			+ 272 Hz	0 %
rotor 4.bending	355 Hz	0 %	- 320 Hz	0 %
			+ 396 Hz	0 %
rotor disk mode	488 Hz	0 %	- 454 Hz	0 %
			+ 538 Hz	0 %
rotor 4.bending	587 Hz	0 %	- 558 Hz	0 %
			+ 631 Hz	0 %

+ forward whirling, -backward whirling
D= damping ratio

C556/001 © IMechE 1999

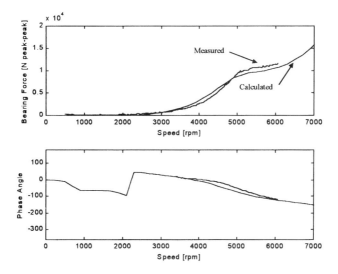

Fig.10 DE Bearing Response to an Unbalance G2 (peak-peak amplitude)

8   CONCLUSIONS

Traditional rotordynamic analysis programs for rotors supported by oil lubricated bearings, are not sufficient to completely predict the rotordynamic behaviour of rotors with active magnetic bearings. For accurate predictions, a comprehensive analysis including the complete bearing control system is necessary. The rotor design analysis, described in this paper, uses a mechatronic part that was programmed in MATLAB. In the design phase of the rotor bearing system, the consequences of changes of the controller or the rotor can easily be evaluated.

For active magnetic bearings in compressors, the axial bearing design usually results in a relatively large disk. The dynamic effects of this disk have to be taken into account.

The agreement between a measured and calculated unbalance response is excellent (see Figure10).

Although the rotordynamic guidelines in the API 617 standard may be applied for magnetic bearing applications, it is felt that these criteria are less suitable for magnetic bearings. Inhouse criteria have been developed. Future issues of the API 617 guideline will be able to benefit from the knowledge of active bearing applications now becoming available.

REFERENCES

[1] Present limits of operation of product lubricated and magnetic bearings in pumps.
M.K. Swann, J. Watkins and K.R. Bornstein, Proceedings from the 14-th international
Pump Users Symposium, 1996.

[2] API 617, Centrifugal Compressors for Petroleum, Chemical, and Gas Service Industries,
6-th edition, February 1995.

[3] Active Magnetic Bearings on Compressor Applications.
J.K. Koo, S.B. Bennett and E.A. Bulanowski, Presentation at the Vibration Institute, May
22, 1986.

[4] Application of Magnetic Bearings and Dry gas Sealing in Axial Inlet process
Compressors. C. W. Pearson, H. J. Aarnink, J. Magee and J.G.H. Derkink, Proceedings
of the 2nd International Symposium on Magnetic Bearings, July 12-14, 1990, Tokyo.

[5] Dynamic Labyrinth Coefficients from a High Pressure Full-scale Test Rig using
Magnetic Bearings. Norbert G. Wagner and Klaus Steff, Proceeding at the 8th Workshop
on Rotordynamic Instability Problems in High Performance Turbomachinery, May 6-8,
1996 Texas A & M University.

[6] Engineering for Rotors Supported on magnetic Bearings. J. Schmied, F. Betschon,
Proceedings of the 6th International Symposium on Magnetic Bearings, August 5-7, 1998,
Boston.

C556/027/99

Benefits of using CFD in compressor design and analysis

P DALBERT
Sulzer Turbo Limited, Zurich, Germany

SYNOPSIS

Over the last 15 to 20 years, computational fluid dynamics (CFD) has become an effective tool in the design of centrifugal impellers and axial compressor blades. The accuracy of CFD modelling has reduced the need for expensive and time-consuming prototype testing. Increased efficiency of the codes and computational power enable multiple stage and blade row calculations. This permits the designer to use CFD in both the design process and in areas where previously only empirical correlations or just a "trial-and-error" method were available, e. g., supporting calculations for root-cause-of-failure analyses. Based on in house experience, this paper describes the benefits and shortcomings of the application of CFD to the design and analysis of industrial turbocompressors.

1 INTRODUCTION

The initial use of CFD in the turbocompressor industry started in the late seventies. During that time effective tools for geometry definition and potential flow computations in two dimensions, with extensions for a simplified blade-to-blade approximation, were developed (throughflow codes). These quasi three-dimensional methods were used successfully through the early eighties. The breakthrough for the turbocompressor industry was initialised with the development of CFD codes for compressible viscous flow calculation at academic institutions, e.g. the Whittle Laboratory of Cambridge University (Denton (1982 [1]).

Computational fluid dynamics for flows relevant to turbomachinery has undergone significant changes since the initial introduction. Early computations of three-dimensional viscous flow in rotating blade rows were extremely time-consuming and rather unreliable due to the coarse grids (which were required to reduce memory and computational time employed). Nowadays, powerful workstations with reliable CFD packages yield accurate results at economical expense. The reduced time for CFD computation, versus prototype testing, has promoted a significant move to more numerical simulations in the industry. In addition, the robustness of the new generation CFD codes, available from universities and commercial code vendors, match the needs of the turbomachinery aerodynamicist. Last, but not least, the multicoloured graphical visualisation techniques applied to CFD simulations help turbocompressor engineers to understand the complex three-dimensional flowfields and to explain them to the management and the customers.

The first part of this paper describes different CFD simulations for design purposes and the second part, the analysis of turbocompressor components.

2   NOTATION

Mu_2	tip speed Mach number
u_2	tip or circumferential speed
$\Phi_1^{\cdot}$	flow coefficient
$\eta_p^{\cdot}$	polytropic efficiency
μ_0	work coefficient
μ_p	pressure coefficient

3   CFD METHODS

Today there is a wide variety of CFD codes available from universities and commercial sources. For the turbomachinery industry, the Dawes BTOB3D codes [2], [3], [4], the NREC code VISIUN [5], CFX-TASCflow by ASC [6] (AEA) and FINE/Turbo by NUMECA [7], to name just a few of them, are commonly in use. They basically all solve the time-dependent Reynolds-averaged Navier-Stokes equations but apply different turbulence models, and use different numerical schemes, to solve the equations. It is beyond the scope of this paper to derive or even discuss the governing equations. The concentration here is on examples of the application of the BTOB3D, FINE/Turbo and TASCflow codes, by engineers of Sulzer Turbo and Sulzer Innotec, to various problems in the field of turbomachinery aerodynamics.

4   CFD APPLICATIONS FOR DESIGN

The primary field for application of CFD in turbomachinery is design. After several years of extensive, and still ongoing, evaluation of the validity (e.g. [8], [9], [10] and more) and limitations (e.g. [11]) of these techniques it is now common practice to integrate them into the design procedure. The three examples below show the utilisation of CFD for design purposes and demonstrate the usefulness of this tool to calculate the extremely complicated three-dimensional flow in turbomachinery.

4.1   Radial compressors

The design of radial turbomachines is usually aimed at optimising the performance in range and power consumption with minimum developmental times and expense. The enormous variety of possible stage geometries, especially in industrial custom-tailored multistage compressors, does not allow the use of CFD to evaluate the best possible stage geometry and combination. On the contrary, an optimisation is carried out for a single stage design and unsuitable geometries are eliminated at the preliminary design phase to reduce the number of possible variations. Therefore, for multistage centrifugal machines it is common practice to use a separate single stage, or even a single impeller development procedure, for the optimisation of the various stages. The single stage performance is then measured in a dedicated test stand and a stage-stacking method is used to design the multistage compressor in a standardised system of geometries and stage performance characteristics (see figure 1).

Initially, after eliminating less promising designs with the preliminary tools, CFD was used in parallel to the throughflow calculations and stress analysis in the development procedure.

But, due to the enormous progress in computing power, and the increasing accuracy of the modelling of the flowfield by modern CFD codes, all 1D, 2D and Quasi-3D methods can currently be regarded as preliminary design tools.

Because modern CFD methods can now be employed economically at the preliminary activities and in post design analysis of stages, the codes are now the most important tool in the design of radial impellers. An example of the use of CFD to design a radial impeller is shown in figure 2. It displays a typical shrouded impeller used in stages of industrial inline compressors. The leading edge of the main blades is relatively far downstream in the radial to axial bend and allows a large shaft diameter. The impeller was designed as a replacement for the second stage impeller (here nominated as "ref.") of Sulzer's inline compressors with integrated coolers and had the following specifications:

- Design flow coefficient $\Phi_{1\,Design}^{*} = 0.105$ (corresponding to $1.17\Phi_{1\,ref}^{*}$)
- Flow Range: $0.07 < \Phi_{1}^{*} < 0.105$
- Required maximum continuous tip speed $u_{2max} = 340$ m/s
- Design tip speed Mach number $Mu_{2} = 0.90$
- Minimum polytropic efficiency $\eta_{p}^{*} > 1.01\eta_{p\,ref}^{*}$
- Pressure coefficient $\mu_{p} = 0.50...0.53$

The imposed geometric constraints for this impeller made the design extremely difficult because the aerodynamic requirements made an adverse effect upon the allowable stress in the blades and shroud disc. Due to high flow rate at high tip speed Mach numbers the leading edge of every second blade had to be moved further downstream. This design modification was in a manner similar to the splitter vanes of inducer type impellers used in single stage compressors. This modification led to high bending stresses in the cover disk and caused the balance of optimisation of aerodynamic performance versus allowable stress levels to be a laborious task. Figure 3 shows the impeller performance characteristics of some of the variants for which BTOB3D computations were performed during the design. The main challenge was to bring into accordance the maximum desired flow rate (at $\Phi_{1}^{*} = 0.12$) and the best point (at $\Phi_{1}^{*} = 0.105$) with an acceptable efficiency and the stress level, since both parameters depend on the blade geometry (length, angle distribution, lean and rake angle etc.).

At the inlet boundary, where the distribution of stagnation pressure, temperature and flow angle has to be defined, uniform conditions can be used. The flow from the return channel is rotationally symmetric and from experience, and CFD computations, it is known that wakes behind the return channel vanes quickly equalise in the accelerated flow around the inlet bend.

As in this example, stage computations were previously seldom used in the design procedure. The main reason for this is that it assumed that the impeller-diffuser interaction is small due to the large distance from trailing edge of the impeller to the leading edge of the vaned diffuser (a factor of 1.15 of the impeller diameter). This holds as long as the matching of impeller and diffuser is determined by prototype measurements with different diffusers. Two examples below will demonstrate that with Fine/Turbo stage calculations will become more and more important.

The use of CFD in the design of radial stages helps to reduce the number of design iterations necessary when prototype measurements show that the goal of a design is not met. Even though the theoretical design needs more time, due to the complexity of the problem, the manufacturing of prototypes and their measurements require typically 5 to 6 months with an increased cost factor of 2 to 3. From this point of view CFD has brought about significant improvements in cost-effectiveness and timescales.

4.2 Axial compressors

The design of industrial axial compressors employed CFD analysis much later than radial machines. The reason for this is described extensively by Casey [12]. The successful prediction of the performance of an axial compressor stage depends on the cascade aerodynamics, the casing and hub boundary layers, the onset of rotating stall and the choking of the blade row. To solve these problems much research effort has been dedicated and nowadays many universal techniques exist for correlating geometrical changes to performance. A design procedure for axial stages is similar to the one shown in fig. 1. The main difference is that in the preliminary one-dimensional study the blading design and the performance prediction of a repeating stage is used for a stage stacking calculation to define the flow channel of a multistage machine. Afterwards the detailed geometry definition is carried out, followed by throughflow, blade-to-blade and three-dimensional computations. The fundamental engineering decisions that determine the ultimate performance potential of a machine are made in the preliminary design process. On the other hand this preliminary design process depends heavily on the available experience and correlations determined with the classical blade profiles (NACA, C4, DCA). When other profiles such as PVD (prescribed velocity distribution) or CD (controlled diffusion) are used, many of the correlations are not valid and another means to determine deviation angles, losses etc., have to be applied. For this reason, more and more three-dimensional calculations are used in the design of axial stages.

Figure 4 shows a BTOB3D computation of the first two stages and the inlet guide vane of a developmental compressor, which has been tested at Sulzer Innotec. In spite of the shortcomings of a calculation procedure, which uses mixing planes between the blade rows, the multi blade row capability of this code reduces the number of iterations that are necessary when an optimisation of the individual blade rows is carried out.

4.3 Inlet volute of an radial compressor

The use of CFD in areas other than impeller, diffuser and axial blading design was delayed for a long time since other components often require a much higher number of grid points. In impellers, diffusers and axial stages the flow is assumed to be rotationally symmetric. This means that only one flow channel with the adjacent blades, and the intermediate splitter blade if necessary, have to be modelled. A volute must be modelled with a computational grid, which covers the whole circumference, while e.g., an inlet casing has at least one line of symmetry. The increased computer power, the availability of cost –effective memory and the optimisation of the codes with multi-grid options, allows the modern designer to model problems with very large grids. For current industrial applications the upper limit of grid points is probably in the order of 300,000 to 500,000 points.

An example of a high grid number model, the inlet volute of a new multistage compressor, is shown in fig. 5. This machine was specially designed for use in the Hydrocarbon Processing Industry (HPI) with the primary design goals of compactness and low manufacturing costs. Such requirements often compromise ideal aerodynamic shaping of the flow channels. Therefore, the inlet volute was designed with the help of CFD. Since the distribution of the flow at the first impeller inlet influences the performance of the first stage and consequently, the complete compressor, the goal of this investigation was to design a geometry with even velocity and flow angle distribution at the leading edge of the impeller. The computations were carried out at Sulzer Innotec with ICEM/CFD for the grid generation and TASCflow for the flow simulation [13]. The CFD solution showed that is possible to find a design, which satisfies both engineering and aerodynamic requirements. Fig. 5a shows the original design, which was similar to a classical casted casing, modified to satisfy the requirements of the welding procedure. The modification resulted in an extremely uneven flow distribution with high velocities in the upper and lower region and a significant mass flow deficit on both sides. In addition, the flow angle variation was more than ±15°. Aided by the CFD solution a new geometry, shown in fig. 5b, enabled the fluid to first distribute around the circumference at low velocity and then be guided smoothly into the radial inlet of the impeller. This aerodynamically improved form is also much easier to manufacture by welding.

5   APPLICATIONS IN THE ANALYSIS

A second increasing field of flow simulations with CFD in the turbocompressor industry is the analysis of problems. Flow problems can occur even when pre-tested standardised stages are used and sometimes, only small deviations from the well-known geometry, or the gas to be compressed, can cause these problems to be severe.

In the first example a case is presented where a standardised stage was put into casing, which was not designed for it. In the second case presented the expected pressure rise of an expander was not delivered and the casing had to be redesigned. The third case presents an assessment of the flow in a secondary flow path, here the leakage flow in the radial impeller-casing gap of a stage, which has a significant effect on the axial thrust in a high-pressure centrifugal compressor. The last example shows the previously underestimated effect of impeller blade cut-back to reduce the pressure rise of a stage.

5.1   Volute

A typical application where CFD can help solve a problem occurred in the revamp of a single stage compressor. The task was to replace the stage (impeller and diffuser) in the existing casing and achieve higher volume flow with lower relative power consumption. This seemed to be a simple problem that could be solved with a newly developed stage for the required mass flow.

After the new impeller was installed the measured polytropic head was 5% lower than predicted near surge, and about 20% lower than predicted near choke. An extensive investigation followed including the stage geometry, the measurement procedure and the geometry of the inlet and the volute. The only possible source for the problem seemed to be in the volute. Therefore the flow from the diffuser exit into the scroll up to the exit flange of the compressor was investigated via a CFD solution. For this investigation the code FINE/Turbo by NUMECA was selected.

A computation of the whole stage would have been too time consuming since the computational grid of inlet, impeller, diffuser and volute would have been much too large. Hence, to determine the inlet conditions of the investigated area a preliminary computation of the stage had to be carried out. This was performed with a basic section of the impeller, from one main blade to the other and a similar section of the vaned diffuser. The flow properties in the section between impeller and diffuser were assumed to be circumferentially mixed out since the two sections do not have the same span and to avoid a necessary unsteady calculation. The results were then used as inlet conditions for the analysis of the volute.

Figure 6 shows on the left-hand side the computed streamlines in the original volute at a mass flow near the design point. The geometry of this volute is unusual because there is a relatively long vaneless part of the diffuser up to the tongue. Also, the tongue has a very low inlet angle. Downstream from the tongue a long channel, around the circumference, follows with a constant width in the axial view but an increasing width perpendicular to this view. At the design point the flow leaves the diffuser at an angle of about 30° (from tangential) and hence, the flow incidence angle to the tongue is very high. This leads to a highly separated flow, which blocks about half of the available channel. At higher flow rates, where the incidence to the tongue is even larger the blockage covers up to two thirds of the channel.

This high amount of blockage leads to high losses, and consequently insufficient polytropic head.

A very simple geometric modification was chosen to solve the problem. A curved metal plate with a leading edge angle of about 30° was added between the vaned diffuser exit and the tongue. As shown on the right hand side of figure 6, this device straightened the flow around the tongue and led to completely uniform flow in the volute without any backflow. The computed increase of polytropic head in the design point recovered about 80% of the previous deficit and about 60% near choke. Since the exit flow of the diffuser is also improved by this solution it is expected that almost the whole deficit can be recovered by this relatively simple geometric modification.

5.2 Expander outlet diffuser

The second example demonstrates a root cause analysis of a problem for an expander turbine used in nitric acid plants.

The problem was a significant shortfall in performance of an expander turbine, here designated Expander 'A'. The expander was designed out of well-known, standardised elements except the outlet casing, which was new. This led to the suspicion of a lack of pressure recovery in the exhaust diffuser and the casing. To correct the lack of performance a redesign of the exhaust diffuser using CFD flow simulations was conducted (designated expander 'B'). The CFX-TASCflow computations were carried out by Sulzer Innotec [14].

Figure 7 shows the two exhaust diffuser geometries in the form of pressure recovery plots in a meridional plane of expanders A (right) and B (left). The plots clearly show that the diffuser of the design A is too short to develop a good pressure rise and that in the exhaust casing an almost complete break down of the pressure rise occurs due to the high flow distortion. The modified geometry shows significantly smoother pressure distribution, which resulted from an improved inflow into the vertical pipe diffuser of the casing. The computed static pressure rise improved by a factor of 8 from casing A to B.

5.3 Swirl brakes

In high pressure compressors flow brakes in the axial gap between impeller shroud disc and casing are used to reduce the axial thrust by changing the pressure distribution along the shroud or to reduce the rotordynamically destabilising effect of leakage flow preswirl into the labyrinth seal. The flow brakes, which can be installed anywhere between the outlet diameter of the impeller and just upstream of the labyrinth seal, have a detrimental effect on the performance of low flow coefficient stages and must be avoided whenever possible. The effect on performance is known from stage measurements in the test stand. The axial thrust, however, is determined by empirical correlations, which depend on the geometry, the pressure level and rise in the stage, the circumferential speed and greatly on the molecular weight of the gas. The thrust has to be minimised due to the capacity of the axial bearing. The residual thrust is the difference of the thrust of the stages and the opposing trust of the balance piston. Because the levels of thrust are large, the correlations applied have to be very accurate.

In order to verify the correlations and to examine the effect of various geometries in the axial gap a series of CFD computations were carried out with CFX TASCflow [15]. Figure 8 shows examples of different geometries for which predictions of the leakage flow in the axial gap were performed. The computations confirmed the empirical constants used in the correlations and render full confidence in the use of the thrust code whenever computations with different geometries are necessary.

C556/027 © IMechE 1999

5.4 Effect of Cut-back

The last example in this review of CFD applications is an evaluation using FINE/Turbo of the effect of cutting back the blades of a centrifugal impeller to reduce the stage pressure rise. During the design of an inline compressor it is often necessary to take into account a certain amount of reserve in rotational speed or pressure rise to overcome uncertainties in the design procedure. Thus, a compressor can be too large and therefore requires correction. A possible measure is to cut-back the impeller blades from the circumference. Since the work coefficient, μ_0, is proportional to the rim speed, u_2^2, the polytropic head is also reduced by the impeller cut-back if the efficiency is not affected.

For many standardised impellers the drop in efficiency which increases the drop in pressure is known from measurements. Sometimes for a stage, which would theoretically be the best suited for a cutback, this effect is not known. It is common knowledge that absolute values of efficiencies are at present not reliably predicted with CFD codes. But, from many comparisons we know that the "Deltas", i.e. the net differences from geometrical changes, are very reliable. The case presented in figure 9 is therefore an ideal example to be analysed by CFD. The question was whether it was possible to cut back the blades of this impeller by 4 to 8% (where the percentage is related to the original impeller and blade diameter) without loosing too much efficiency and pressure. As shown on the sketch in figure 9, the computations were carried out using the stage geometry with a radial inlet to the impeller and a vaned diffuser until the half the bend of the return channel. This configuration was also necessary to investigate the effect of the changing impeller outflow angle on the diffuser performance. The curves show on top the isentropic efficiency and below the corresponding pressure coefficient for the base geometry (0), a 2% (1), a 4% (2) and an 8% (3) cutback of the impeller blades.

The result was sobering. The range of the stage decreases due to the increase of the absolute flow angle into the diffuser and the peak efficiency drops and moves to the left. The efficiency drop starts with a cutback of 2% and is 7 points lower than the base value at a flow coefficient of 0.1. At the same operating point the pressure coefficient is reduced by more than 20%. These calculations show that a cutback of this impeller would have been unexpectedly disastrous.

6 CONCLUSIONS

In a series of examples the use of CFD flow simulations in the design and analysis of turbomachinery components was shown. These examples showed that computational fluid dynamic tools have given the aerodynamicist an irreplaceable tool in the design, development and understanding of the extremely complicated three-dimensional flow in axial and radial turbomachines. CFD will never replace component and machine testing but it already helps to better understand the flow physics and it is a most effective tool to reduce the extent of expensive and time-consuming experiments. Because of the inherent numerical and modelling errors, the methods have to be validated prior to their use. In addition, it is important to realise that only experienced engineers, who have enough knowledge of turbocompressor aerodynamics and the limitations of their tools, are able to assess the results and put into practice the benefits of these highly sophisticated tools.

ACKNOWLEDGEMENTS

The author would like to thank his employer for permission to publish this paper and to use the data presented. He would particularly like to thank all the colleagues at Sulzer Turbo and Sulzer Innotec who carried out all the work summarised in this paper.

REFERENCES

[1] Denton, J. D., 1982, "An improved time marching method for turbomachinery flow calculation", ASME paper 82-GT-239, 1982

[2] Dawes, W. N.: "Development of a 3D Navier-Stokes solver for application to all types of turbomachinery", ASME paper 88-GT-70, 1988.

[3] Dawes, W. N.: "The simulation of three-dimensional viscous flow in turbomachinery geometries using a solution adaptive mesh methodology", ASME Paper 91-GT-124, 1991.

[4] Dawes, W. N.: "Toward improved throughflow capability: The Use of three-dimensional viscous flow solvers in a multistage environment", Trans. of ASME, Journal of Turbomachinery, Vol. 114, Jan. 1992.

[5] Northern Research and Engineering Corporation, "Simultaneous three-dimensional CFD-analysis of rotating and stationary bladed passages in turbomachinery, a computational system", Massachusetts, 1995.

[6] CFX-TASCflow Documentation, Version 27, Advanced Scientific Computing LTD., Waterloo, Ontario, Canada, 1998.

[7] FINE™, User Manual Version 3.0, Numeca International, 1997.

[8] Casey M. V., Dalbert, P., Roth, P.: "The use of 3D viscous flow Calculations in the design and analysis of industrial centrifugal compressors", Trans. of ASME, Journal of Turbomachinery, Vol.114, Jan. 1992

[9] Dalbert, P., Wiss, D.: "Numerical transonic flow filed predictions for NASA compressor rotor 37", ASME Paper 95-GT-325, 1995.

[10] Eisenlohr, G. et al., "Analysis of the transonic flow at the inlet of a high pressure ratio centrifugal impeller", ASME Paper 98-GT-24, 1998.

[11] Casey, M. V.: "The industrial use of CFD in the design of turbomachinery", in Turbomachinery design using CFD: AGARD Lecture Series 195, 1994

[12] Casey, M. V.: "Computational methods for preliminary design and geometry definition in turbomachinery", in Turbomachinery design using CFD: AGARD Lecture Series 195, 1994

[13] Holbein, P.: "Designverbesserung von HPI-Einlaufgehaeusen mittels CFD", Internal Report, STT.TB98.025, October 1998.

[14] Muggli, F., Sieverding, F.: "Simulations of the flow in the exhaust casings of the expanders …", Internal Report, STT.TB98.019, September 1998.

[15] Mack, P.: "Numerische Strömungssimulation des deckscheibenseitigen Radseitenraumes einer Radialverdichterstufe mit Statorstreifen", Internal Report, STT.TB96.018, May 1996.

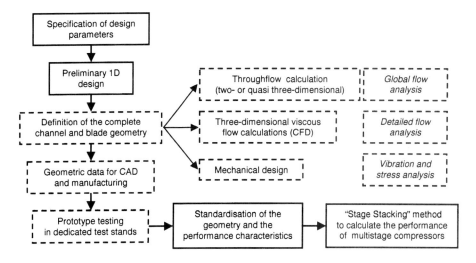

Figure 1: Design procedure for radial stages

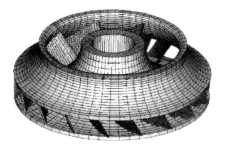

Figure 2: Typical centrifugal impeller of an inline multistage compressor

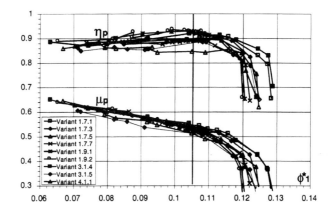

Figure 3: Computed performance characteristics during design

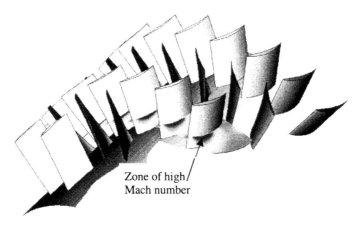

Zone of high
Mach number

Figure 4: Mach number distribution in the first stator of an multi-blade-row CFD
computation in an axial compressor.

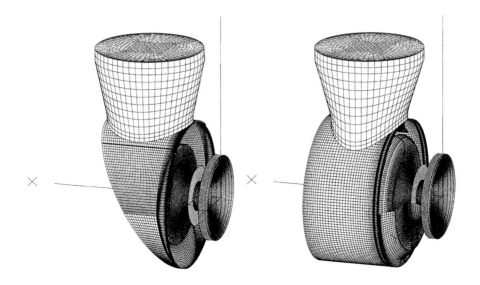

Figure 5: Inlet volute of an centrifugal compressor

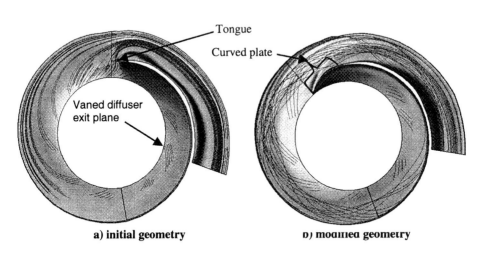

a) initial geometry   b) modified geometry

Figure 6: Original and modified volute

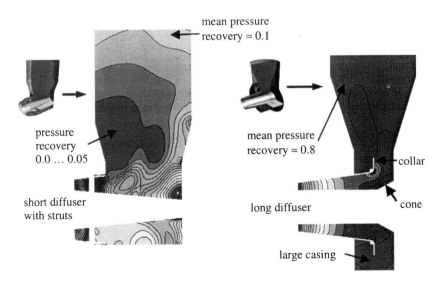

mean pressure
recovery ≈ 0.1

pressure
recovery
0.0 ... 0.05

short diffuser
with struts

mean pressure
recovery ≈ 0.8

collar

long diffuser

cone

large casing

Figure 7: Flow in expander outlet A (left) and B (right)

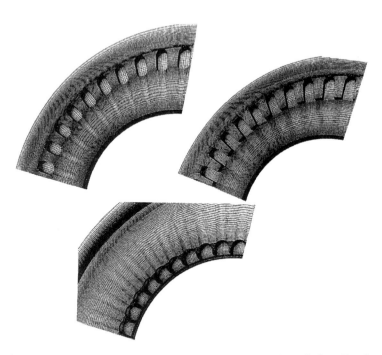

Figure 8: Flow brakes in the axial gap between shroud and casing of a low flow impeller

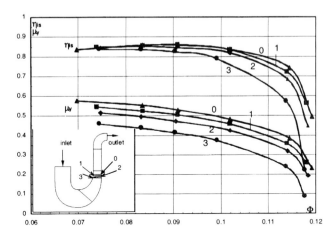

Figure 9: Effect of cut-back on efficiency and pressure coefficient

C556/005/99

Modal parameter estimation of rotating machinery

K POTTIE, J G A MATTHIJSSEN, and **C J J NORBART**
FLowserve BV, Etten-Leur*, The Netherlands
L J P GIELEN
LMS International Engineering Services, Leuven, Belgium

SYNOPSIS

Experimental modal analysis and testing of rotating machines imposes a significant complexity, compared to the approach for passive structures. In order to evaluate the applicability of modal analysis to rotors in operation, two commercially available algorithms were adapted to match with specific rotor behaviour. Both methods were validated by means of a numerical experiment. Then the modal analysis techniques were applied on a boilerfeed test pump with magnetic bearings. The diagnostic capabilities to detect increasing clearances were investigated. Additional knowledge of the rotor system allows to obtain some non-identifiable modal parameters or some system parameters. Finally a portable electromagnetic exciter was built; an experimental modal analysis was performed on a booster pump and on a boilerfeed pump in a thermal power plant.

INTRODUCTION

Experimental modal analysis and modal testing have become a popular method for modelling the dynamic behaviour of a wide variety of passive structures (1), by means of three groups of easy-to-interprete modal parameters:

(a) Eigenfrequencies and damping ratios. Both define the poles of the rotor.

(b) Modeshapes, which describe the topology of the standing waves in the rotor at resonance. Mathematically the modeshapes correspond to the right eigenvectors of the system matrix.

*Previously BW/IP International BV

(c) Modal participation vectors, which define, for each mode, the contribution of a unity force, applied in each degree of freedom (DOF), to the frequency response of the system. Mathematically the participation vectors correspond to the left eigenvectors of the system matrix. For passive structures, left and right eigenvectors are identical.

Application of modal testing for parameter identification and diagnostics of rotating machines has several specific aspects compared to passive structures. Typically this concerns (2):

(a) Non-symmetric and cross-coupled stiffness, damping and mass matrices. These matrices represent the relation between some specific forces/moments and the lateral/angular vibrations of the rotor:
   - Gyroscopic effects.
   - Fluid interaction between the rotating and the stationary parts. This interaction is generated by and dependent upon vibrations of rotor or stator. It occurs in bearings, wearings and more generally in all locations with confined clearances.
     If the vibrations of rotor/stator are small, they can be considered as perturbations of the equilibrium position of the rotor. The fluid interactions can then be linearized ; the constant matrix coefficients resulting from this linearization are referred to as motion dependent interaction (MDI) effects.

(b) Due to the structure of the system matrix (particularly if cross-coupling and damping are important), the reciprocity principle is not valid anymore. Consequently one should excite the rotor in all measurement locations to obtain experimentally the modal participation factors.

(c) Non-proportional damping.

(d) MDI effects, gyroscopics and damping introduce sets of forward, backward and even mixed precession modes. As these modes are often closely coupled, a real-life structure requires the use of global parameter estimation schemes.

(e) As the MDI coefficients depend upon speed and other operational parameters (such as: pressures produced by the turbomachine, thermal effects and foundation deformations), modal testing should occur during operation.

(f) The identification techniques must be capable of handling noise. In this case, noise accounts for all deviations from the ideal model, which may be due to unmeasured inputs, small non-linearities, non-stationary effects, and instrument noise. The unmeasured inputs are particularly important since each rotor is subjected to inaccessible hydraulic excitation forces. These can be sinusoidal (e.g. vane-passing frequencies or acoustic resonances) or broad band (e.g. recirculation, cavitation, and fluid instabilities). Most of these unmeasured inputs are generated by the pumping action of the impellers.

PARAMETER EXTRACTION METHODS

Two methods were modified to calculate global modal parameter estimates for rotors : the polyreference Least Squares Complex Exponential (LSCE) and the Frequency Domain Direct Parameter Identification (FDPI).

The LSCE method is a time-domain method. The measured frequency response function (FRF) matrix is inverse Fourier transferred to an impulse response matrix. Next, these impulse responses are approximated by a sum of damped complex exponential (thus sine and cosine) functions. Negative damping coefficients are hereby allowed. The number of these damped harmonic functions, corresponding to the model order, is gradually increased. When the model order is equal to or higher than the number of resonances, the estimates of parameters associated with each damped harmonic function will stabilize. The LSCE method yields poles and participation factors only. The modeshapes are calculated in a second step, response DOF by response DOF, using a least squares fit of each individual transferfunction in the frequency domain (LSFD).

As the LSCE method is a time-domain method, it's major drawback is related to the poor estimation of highly damped modes. Their contribution to the impulse response will die out after a few time samples. The residual error of the damped harmonic function fitted to this distribution might be quite important. In case of lightly damped structures, however, the LSCE method is considered as the reference method for the evaluation of other methods.

The FDPI method identifies a low order complete direct model from multiple output FRF data. It generates global estimates for system poles, participation factors and modeshapes. This implies that the LSFD modeshape estimation is not necessary. In practice however, curvefitting every FRF by LSFD, using poles and participation factors estimated by FDPI, generally improves the accuracy of the estimated modeshapes.

The FDPI method is a frequency domain method. In case of lightly damped structures, the contribution of a single mode to the FRF spectrum is restricted to a few spectral lines. The pole estimation process will suffer from a lack of overdetermination. The higher the damping, the higher the overdetermination in the frequency domain. This implies that highly damped poles will be estimated in a very accurate way.

NUMERICAL EXPERIMENT

The aim of this phase was to check whether the modified LSCE and FDPI algorithms are able to estimate the modal parameters from a set of measured FRF, without having to struggle with noise effects, nor with the limited amount of DOF available for artificial excitation, nor with a lack of observability. The numerical experiment consisted of five steps:

(1) An analytical forced response analysis was performed for two rotors, by means of a commercial finite element (FE) program, in the frequency band 5-200 Hz :
   - Rotor at 5081 rpm. This machine is characterized by a mix of lightly damped modes and some much heavier damped modes. The model is characteristic for boilerfeed pumps, and to some extent for turbines and compressors.
   - Highly damped rotor, also at 5081 rpm. The model is relevant for a 7-stage horizontal pump transporting high viscosity fluids, with a rated flow of 411 m³/hr and a corresponding TDH (total dynamic head) of 2566 m. The rotor is supported by hydrodynamic bearings.

It was assumed that the radial rotor vibrations are small, so that the structure-fluid interaction could be linearized by MDI-matrices (calculated by a CFD code). It was further assumed that the displacements to unity forces are available at 10 measurement locations in two lateral (Y and Z) directions. These locations are spread over the rotor. It was finally assumed that all measurement locations are also accessible for artificial force excitation.

(2) The results files (embedded in the database structure of the FE code) were accessed by an external postprocessor, without any format conversion and/or data reorganization. The calculated forced responses were stored as a matrix of FRF, in the database of the data-acquisition software. This FE postprocessing reduced the original FE output data size by a factor of 300, and resulted in an equivalent experimental database for validation of parameter extraction algorithms.

(3) Application of LSCE and FDPI. To determine the optimal number of modes, the frequency and damping estimates are compared, for experimental models with an increasing amount of modes (8). The basic idea is that estimates of frequency and damping corresponding to physically occurring modes reoccur as the number of modes is increased, whereas computational (phantom) poles will appear and disappear in a random way with changing model order. A stabilization diagram presenting the evolution of frequency and damping as the number of modes is increased, is shown in figs 1 and 2, for a global (complex) estimation based on a 20x20 FRF matrix, in the frequency band from 7 to 198 Hz. The subsequently estimated poles are plotted on top of the amplitude of the "Sumblocks" function. The real part of the sum blocks corresponds to the sum of the absolute values of the real parts of all FRF. The imaginary part of the sum blocks corresponds to the sum of the absolute values of the imaginary parts of all FRF. Both the amplitude and the phase of this summed spectrum provide useful information to visually identify resonance frequencies.

Fig 3 presents hardcopy snapshots of one experimentally identified modeshape (first-bending-like mode, forward whirl) of the highly damped rotor.

From the stabilization charts, it was concluded that FDPI suffers less from phantom poles than LSCE. This indicates that the amount of damping in the rotor system requires a frequency domain parameter estimation method. Further on, the combination of a frequency close to the upper frequency limit and a relatively high damping value in one pole made identification using the LSCE method impossible.

(4) Prior to the actual comparison of the FE and the equivalent experimental modal models, a topological data integration step was necessary so that a genuine comparison of the results could be made. It involved the access of both models from one single software platform, and the transformation of the geometry of the FE model (plus reduction of DOF) to match the topological definition of the experimental model. The result of this transformation is a table with corresponding node pairs.

(5) Finally the FE and the experimental modeshapes could be compared. This was done via visual comparison of the modeshapes (fig 3) and via a matrix with MAC (modal assurance criterion) values. The MAC value between any experimental and any analytical vector expresses the degree of likelihood between both. A value of 1 indicates a perfect

correlation, a value of >0.7 indicates a good correlation, while a value < 0.5 indicates poor correlation.

Table 1 (lightly damped rotor model) and table 2 (highly damped rotor model) contain the poles identified by the experimental analysis, the pole definition of the FE normal modes solution, the absolute difference between both in frequency (Hz) and damping (% of critical damping) and the MAC value between the corresponding modeshapes.

From this verification phase, it was concluded that FDPI and to a lesser extent LSCE are suitable to estimate the modal parameters that characterize the dynamic behaviour of any rotor.

MEASUREMENT ON A BOILERFEED TEST PUMP

The LSCE and FDPI identification techniques were then applied to identify the modal parameters of a (destaged from 7 to 2 central impellers) boilerfeed pump (Fig. 5). The pump was equipped with active magnetic bearings (AMB). These do not only suspend the rotor, but they also allow to impose artificial excitation forces.

Sensors

The pump was instrumented with several probes, namely (Fig. 4) :
(a) Shaft proximity probes. Probes TM1A-TM2A and TM1B-TM2B were at both sides of the outboard magnetic bearing. TM3A-TM4A and TM3B-TM4B were at both sides of the inboard AMB. Dymc. 1-2 were installed at the thrust disk, Dymc. 3-4 at the middle of the center-stage piece between both impellers, Dymc. 5-6 at the coupling flange. Probes MTRX1-MTRX2 monitored shaft vibrations at the location of the fifth impeller. MTRX3-MTRX4 were installed at the location of the first impeller.

(b) Two sets of sensors that allowed to measure all forces: (1) currents in and displacements (TM-probes) at the coils of the AMB; (2) strain gauge load cells.

(c) 3D accelerometers on the top corners of the pedestals as well as on the top line of the outboard bearing bracket / barrel / inboard bearing bracket. These accelerometers enabled measurement of the responses of the static structure due to the electromagnetic excitation forces on the rotor.

(d) Compensation accelerometers at the load cells, to take into account the inertia forces and moments of the AMB stators.

(e) Compensation accelerometers on the bearing housings, to correct the measurements by the proximity probes.

Force calibration

Each radial magnetic bearing consists of two diametrically opposed coils. The resultant load in Y or Z direction is described by

$$F_{net} = K_1 \left(\frac{I_{01} + \Delta I_1}{\varepsilon_{01} + \Delta \varepsilon} \right)^{\ell 1} - K_2 \left(\frac{I_{02} + \Delta I_2}{\varepsilon_{02} + \Delta \varepsilon} \right)^{\ell 2}$$

Bearing currents I_1, I_2 and air gaps ε_1, ε_2 contain two contributions. The first components are related to the position of the non-excited shaft. The second components Δ describe the change due to artificial excitation of the rotor. K_1, K_2, l_1 and l_2 depend upon the individual properties of each coil and control loop ; they had to be calibrated.

Therefore, the non-rotating and dry shaft was loaded by tuneable stiff ground connections, in which the forces were measured by precision load cells. Equilibrium equations allowed to verify the AMB forces. Three major errors appeared when using the force model (3).

-   Dependency of calculated force output on static rotor position.
-   Dependency of calculated force output on the amplitude of the load. The error is linear for low loads and non-linear for high loads.
-   Hysteresis effects.

To cope with some of the force errors, a large number of external load conditions were applied for a number of different shaft positions, resulting in a force error matrix. The latter provided design data for a hardware correction, which allows a tuneable linear adjustment of force output with the original force output value, the position information and also a DC offset value, independent for each bearing degree of freedom.

As all previous corrections were based on DC calibration, another series of measurements was made to investigate the accuracy of the AC force output:

-   Linearity check: AMB-FRF were measured for different input levels, with a max./min. ratio higher than 4. All FRF are very similar.
-   Comparison of measured FRF by hammer excitation (performed on the rotor while suspended in flexible slings) and calculated FRF (FE model, tuned to match with the test in slings) on one hand, with the measured FRF by stepped sine AMB excitation on the other hand (performed on the dry and non rotating rotor). Important deviations were found (Fig. 6a) between analytical model and measurement, particularly for frequencies below 30 Hz. Further on, the AMB introduce additional structural damping in the rotor.
-   Mass identification from radial FRF by AMB excitation was impossible because of the flexibility of the rotor.

In order to improve the quality of the FRF at frequencies below 30 Hz, a set of 8 strain gauge load cells was installed between each AMB stator and the casing of the pump (Hall sensors were not feasible for this testrig). The load cells are diametrically opposed and connected in anti-parallel, so that cross-coupled effects between the lateral directions are eliminated. The measured strain gauge signals were compensated by the inertia effects of the AMB stators. Fig. 6b shows that the experimental FRF were improved at low frequencies.

However, the implementation of the load cells introduced a flexible connection between the pump casing and the AMB. It turned out that the AMB's generate considerable magnetic and inertia moments on the rotor, which are highly correlated with the magnetic forces. It was impossible to eliminate the casing dynamics from the FRF obtained on the rotor (as should be, since the dry rotor can theoretically be considered as a free body on which excitation forces are applied), particularly from 160 Hz on.

From all these tests, it was concluded that the accuracy of the FRF on the dry rotor was fairly good up to 160 Hz. Nevertheless, the dynamic accuracy of the measured forces remained significantly lower than the (static) values mentioned in (4).

Shaft excitation methods

The shaft was excited with stepped sine excitation forces. As for broadband excitation, the approach consists of the accumulation of auto- and cross power spectra of all displacement signals and the excitation forces, followed by estimation of the corresponding set of FRF.

For sinusoidal excitation the different spectra are built frequency per frequency. At each frequency, new time histories are measured and the fundamental signal component corresponding to the excitation frequency is retained. Although the stepped sine approach is considerably slower than broad band testing, the major advantage is that an important noise reduction is obtained.

When multiple inputs are used, these must be made uncorrelated to allow correct calculation of FRF with respect to all inputs. Therefore, 6 different force combinations were used to generate the FRF (3).

Test results

Fig. 7 shows a typical FRF obtained for the pump rotating at 2500 rpm, at best efficiency point (Flowrate 340 m^3/hr, TDH = 198 m), with medium pump clearances. There are important differences between the dry non-rotating shaft and the operational ("wet") machine:

- Additional resonances below 35 Hz. These correspond to two pairs of rigid body modes (MAC numbers between the identified modes and the rigid body modes fitted on the identified are higher than 0.97). Each modepair consists of one backward whirling (BW) mode and of one forward whirling (FW) mode, fig. 8. The rigid body modes are caused by the MDI effects, which introduce additional bearings along the rotor. Contrary to the dry and non-rotating shaft, which is only suspended by the forces of the AMB, the wet rotor can not be considered as a free body anymore.

- The forward whirling mode of 14.1 Hz is negatively damped. In practice, however, the rotor remained stable since it is part of a mechatronic system. The control loops of the AMB's stabilized the rotor plus MDI bearings.

- The first bending mode at 54.0 Hz (identified during the test in flexible slings) is split into one highly damped BW mode at 50.8 Hz and into one FW mode at 59.9 Hz. Numerical simulations revealed that gyroscopics have only a small contribution to the separation of BW & FW modes. For this pump, the separation is essentially caused by the viscous damping (Fig 8) MDI coefficients.

- The second bending mode at 175 Hz (test in flexible slings) is split into a BW mode at 136.7 Hz and a FW mode at 147.7 Hz (Fig. 8). Numerical simulations showed that the "direct" angular stiffness and inertia MDI coefficients at the centre stage impellers are essential for this mode (The numerical model with only radial MDI coefficients estimates the second bending BW and FW modes at 144 respectively 155 Hz). The damping ratios of both modes are very low and suggest that the rotor is nearly unstable (Numerical predictions are about 1.5 %). It was impossible to figure out whether these damping ratios are underestimated by the applied analysis techniques.

Besides these pure rotor modes, two coupled rotor-casing modes were identified (Fig. 9). The first at 66.1 Hz shows that the entire pump is twisting around a vertical axis. The shaft is entrained. The first FW bending mode is still visible in this modeshape. The second coupled mode at 82.1 Hz shows that the pump barrel is rocking on the pedestals. Once again, the shaft is entrained. The accuracy of the last pole is somewhat doubtful, as there was a range from 76 Hz to 83 Hz for which all modal vectors were highly correlated.

No other modes were identified with this set of excitation forces and measurement locations. Synthesis of FRF from the modal parameters and a subsequent experimental modal analysis on the structure of the machine (without the pump running, impacting on other locations) revealed that additional structural modes do occur:
- bending mode of the skid, at 34 Hz
- rotation of the suction piping and pump around an axis parallel to the pump shaft, at 138 Hz.
- torsion of skid, at 146 Hz.

Table 3 lists the summed modal participation factors (inclusive scaling factor) for each mode and for each DOF input. The relative importancy of each mode in the frequency band 0-150 Hz is tabulated in the last column, for all 4 applied input forces. The values must be interpreted with uttermost care, since they do not reflect the effect of important forces at other locations, for instance at the impellers. From table 3, it can be observed that the first 6 modes were well excited by the inputs. The modal participation factors for the second bending modepair are low ; they could partially explain why it was possible to excite these modes without damaging the pump.

As long as it is impossible to excite the rotor at all locations where important forces occur, the remaining unknown participation factors should be obtained in another way. For instance :
(a) By means of FE models, which ideally should be updated, via an interface, with the results of the experimental modal analysis. Such interface already exists for passive structures.
(b) Assuming that the cross-coupled effects are dominated by gyroscopics and that damping is small (5), the participation factors can be readily derived from the modal vectors.

All experimental results were obtained by combining LSCE and FDPI. For this test case, LSCE was used to identify the poles in one leap, over a wide frequency range (6-200 Hz). As it failed to set a pole at the peaks of the "Sumblocks" function (cf. Stabilization diagram, fig.

                                              C556/005 © IMechE 1999

10) that correspond to the rigid body modes, LSCE had to be assisted by FDPI. FDPI was also used to verify the other modes, with emphasis on the highly damped ones.

Condition monitoring capacity

To investigate it's monitoring capabilities, an experimental modal analysis was applied on a pump with increased wearring clearances. The clearance increments correspond to a wear pattern that is determined by the first bending modeshape (table 4).

It was observed that the rigid body modepairs and the first bending modepair are sensitive to variations of the MDI coefficients (table 5). The first bending modes are determined by :

(a) Direct-coupled radial stiffness coefficients of the MDI matrices. They consist of two contributions (6) :
   (1) One positive term, caused a by non-symmetric circumferential pressure distribution at the inlet of the wearrings (Lomakin effect).
   (2) One negative and destabilizing term, caused by centrifugal forces acting upon the rotating fluid. For this particular rotor, the overall stiffness values decrease with increasing clearances, resulting in smaller values of the eigenfrequencies of the first bending modes.
(b) All other radial MDI coefficients, which decrease with increasing clearances (the rates of decrement are different).

Simulations on other pump rotors show that increasing wearing clearances can reduce the centrifugal effect much more than the Lomakin effect, particularly for fastly rotating machines. In such case, the overall stiffness and the eigenfrequencies could increase, possibly after having decreased, whereas the clearances grow continuously.

Further on, once the clearances become high, the changes of the MDI effects become so small that the rotordynamic behaviour only varies marginally, imposing a need for extremely accurate FRF.

The second bending modeshapes appear to be insensitive to variations of the clearances of the wearrings. This might be because the dominating MDI coefficients are not caused in the deteriorating wearrings only, but to an important extent at the (unchanged) impeller shrouds as well. Further on, the effect of the angular coefficients can be observed only indirectly (that is: via radial displacements), which makes accurate identification more difficult.

PORTABLE FIELD TEST EXCITER

Set up

A portable vertically split, non-contacting electromagnetic exciter was designed and built. The exciter was applied on two different machines in a thermal powerplant (design early 70's):

(1) A horizontal motor-driven 6-stage boilerfeed pump with a rated flow of 500 m³/hr, 2155 m TDH, 6300 rpm, 2.95 MW, hydrodynamic bearings. The exciter was mounted over the coupling flange of the pump (Fig.11).

(2) A single stage booster pump with a rated flow of 500 m³/hr, 52 m TDH, 1495 rpm, 70 kW, hydrodynamic bearings . The exciter was mounted over a special hub at the drive-end of the pump.

Each time, a laminated bushing allowed to : (a) match the bore of the exciter with the outer diameter of the rotor (nominal radial gap: 0.75 mm). (b) apply high force levels without excessive heating of the rotor. The exciter was equipped with proximity probes and force probes. No additional proximitors could be installed on the main pump. Two extra proximitors were installed at outboard site of the booster pump.

Test results

Fig 12 shows a wireframe model of the exciter and of the centre of the shaft. Fig. 13 presents the BW and FW "modeshapes" (plus corresponding frequencies/damping ratios), obtained on the boilerfeed pump, in a frequency range 5-150 Hz. No other rotor modes were detected.

Comments

Despite some FRF of good quality were generated and some modal parameters could be extracted, this prototype exciter test revealed some practical problems, specifically for trouble shooting:

- The rotor must be stopped to allow for mounting and dismounting the exciter.
- A laminated hub or coupling with a laminated bushing must be mounted on the rotor. Besides a lot of engineering and installation effort, this also changes the rotor dynamic behaviour of the shaft. This problem can be avoided by eliminating the laminated hub (reducing the load capacity of the exciter) or by a permanent installation of the hub (so that it becomes part of the rotor).
- Alignment of the exciter must be done carefully, by skilled personnel. This action can last ½ day.
- The electromagnetic exciter, it's support and the baseplate on which the support is mounted should be sufficiently stiff (for the baseplate, this implies correct crouting and welding). A stiff construction will guarantee that the electromagnetic forces are mainly transferred into vibrations of the shaft. This stiffness should also be adjustable, in order to separate pump resonances from exciter / support resonances.
- There should be enough space to accommodate the support.
- Sufficient locations for installing proximity probes on the shaft and for compensation accelerometers are necessary to extract maximal modal information. Proximity probes shall not be used without compensation accelerometers.
- Calibration of the force transducers requires the construction of a set of calibration rotors, at the manufacturer's works.

C556/005 © IMechE 1999

Most of these problems can be circumpassed if the rotor is initially designed and built to access the portable exciter. An alternative can be to adapt the machinetrain during shutdown for maintenance/repair/ refueling.

CONCLUSIONS

- Two experimental modal analysis techniques, LSCE and FDPI, were applied on :
  - FRF generated by FE models of pump rotors. The FE models were postprocessed and linked with the data acquisition platform.
  - FRF obtained on a boilerfeed pump equipped with active magnetic bearings. The excitation forces were generated by the bearings.

  Eigenfrequencies, damping ratios and both backward as forward whirling modeshapes were identified. Modal participation factors were identified in those locations where excitation forces were applied. As such, LSCE and FDPI have the potential to characterize the dynamic behaviour of any linear(ized) rotor. They can be applied to assess the structural integrity of the rotor during the prototyping phase, acceptance testing or trouble shooting.

- With the same sets of excitation forces, it was possible to perform a modal analysis on the rotor and on the stator. This allowed to make a distinction between rotor modes and some casing modes, and to observe the coupling between casing and rotor. Such coupling occurs, for example, if the rotor excites a resonance of the bearing brackets.

- Both FDPI as LSCE can identify correctly the modal parameters of rotors. However, LSCE should be assisted by FDPI in the case of : coexisting lowly and highly damped modes, poles that are close to the upper excitation limit, rigid body modes.

- Experimentally identified modal parameters are sensitive to variations of the MDI coefficients of the wearrings in pumps. The eigenfrequencies and the damping ratios change due to increased clearances, caused by wear.

  As such experimental modal analysis shows some potential to be used as a condition monitoring and diagnostic tool. It provides a compromise between passive vibration monitoring and full disassembly of the machine. Since not all modes are equally sensitive to the effect of e.g. wear, the diagnostic capability increases with the excitation over a wide frequency range with a maximum amount of eigenfrequencies.

- It must be kept in mind that all identified modal parameters are a mathematical condensation of much more physical rotor parameters. The diagnostic capability of experimental modal analysis is further increased if the experiment is supported by a theoretical rotor model. Simulations can then be used to study the effect of changing MDI coefficients (e.g. caused by wear) on the modal parameters.

- A portable exciter was built and applied on a booster pump and on a boilerfeed pump in a thermal power plant. Both pumps were equipped with hydrodynamic bearings. A set of modal parameters was identified. The exploitation of the capabilities of the in-the-field excitation for modal parameter estimation on existing and operational machines still requires to tune practical installation procedures, but is proven to be suited for rotating machinery diagnostics.

ACKNOWLEDGEMENTS

The research reported in this paper has been made possible by financial support of the European Commission in the framework of the Brite-Euram II program, contract BRE2-CT94-0945 (7), (8).

The authors also express their gratitude to the project partners Glacier (portable exciter) and Iberdrola (pumps in thermal powerplant) for their contribution to this paper.

REFERENCES

(1) W. Heylen, P.Sas, S. Lammens. *Modal   Analysis, Theory and Testing.* 18th ISMA Conference, 1993, Leuven, Belgium.

(2) R. Nordmann. *Schwingungberechnung von nichtkonservativen Rotoren mit Hilfe von Links- und Rechts-Eigenvektoren*, VDI-Berichte Nr. 269, 1976.

(3) K. Pottie, G. Wallays, J. Verhoeven, R. Sperry, L. Gielen, D. De Vis, T. Neumer, M. Matros, R. Jayawant. *Active magnetic bearings used in BW/IP centrifugal pumps.* 4th International Symposium on Magnetic Bearings, August 1994, Zurich, Switzerland.

(4) P. Förch, C. Gähler, R. Nordmann. *AMB System for rotordynamic experiments: calibration, results and control.* 5th International Symposium on Magnetic Bearings, August 1995, Kanazawa, Japan.

(5) I. Bucher, D.J. Ewins, D.A. Robb. *Modal testing of rotating structures: difficulties, assumptions, and practical approach.* 4th International Conference on Vibrations in Rotating Machinery, I MechE Conference Transaction 1996-6.

(6) H.F. Black, D.N. Jenssen. *Dynamic hybrid bearing characteristics of annular controlled leakage seals.* Journal of Mechanical Engineering, 184, 92-100, 1970.

(7) J.J. Verhoeven. *Advanced and preventive diagnostics of fluid handling machinery using artificial rotor excitation and dynamic parameter identification techniques (BE-7356),* The 13th European Maintenance Conference and Fair, May 1996, Copenhagen, Denmark

(8) *LMS Cada-X User Manual, Modal Analysis, Rev 3.4.*

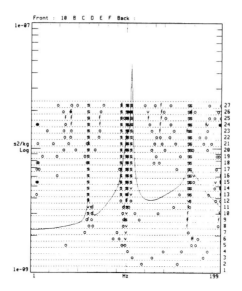

Fig 1: LSCE Stabilisation diagram and Sumblocks function for low damped rotor.
Horizontal lines correspond with the number of modes of the mathematical model ; o =
unstable pole ; f = frequency of the pole does not change within the tolerances ; d =
damping of the pole does not change within the tolerances ; v = pole vector does not
change within the tolerances ; s = frequency, damping and vector are stable within the
tolerances.

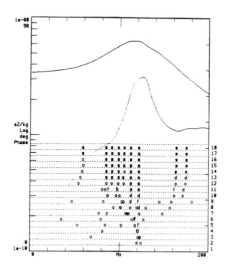

Fig 2: FDPI Stabilisation diagram and Sumblocks function (Upper curve : amplitude.
Lower curve : phase angle) for highly damped rotor.

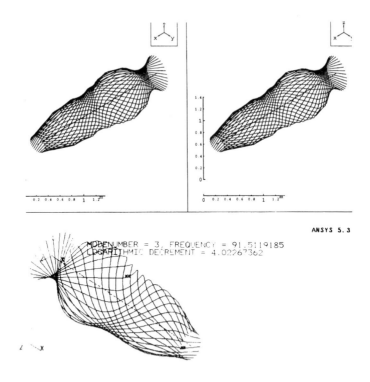

ANSYS 5.3

Fig 3: First–bending-like mode of highly damped rotor (Left top : CADA-X FDPI curve, based on the FE response vectors in 10 measurement locations. Right top : modal model from the FE normal modes solution, after reduction of DOF and after parsing of FE and experimental geometry. Bottom : modal model from original FE normal modes solution, with full amount of DOF.)

CADA -X Frequency	CADA-X Damp (%)	ANSYS Frequency	ANSYS Damp (%)	Freq. Diff. (Hz)	Damp. Diff. (%)	MAC (/)
60.29	20.69	60.13	20.83	-0.16	0.14	1.00
60.73	2.30	60.85	2.30	0.12	0.00	1.00
94.61	15.47	94.02	14.89	-0.60	-0.58	1.00
100.23	2.56	100.68	2.26	0.45	-0.30	0.99
101.57	15.72	100.87	16.23	-0.70	0.51	1.00
105.34	-0.32	106.14	0.03	0.80	0.35	1.00
164.99	4.46	163.40	4.53	-1.59	0.07	0.97
167.23	6.65	168.78	6.70	1.55	0.05	1.00
184.01	32.90	184.33	32.89	0.32	-0.01	1.00
189.92	40.35	189.47	40.37	-0.44	0.02	1.00

Table 1: Mode pair table for lowly damped rotor model.

CADA -X Frequency	CADA-X Damp (%)	ANSYS Frequency	ANSYS Damp (%)	Freq. Diff. (Hz)	Damp. Diff. (%)	MAC (/)
58.95	90.25	58.99	90.24	0.05	-0.01	1.00
85.18	72.22	85.18	72.22	0.00	0.00	1.00
91.54	53.90	91.49	53.94	-0.05	0.04	1.00
98.86	45.15	98.89	45.14	0.02	-0.01	1.00
105.17	82.13	105.39	82.10	0.23	0.03	1.00
113.14	26.97	113.14	26.97	0.01	0.00	1.00
121.97	25.14	121.95	25.14	-0.02	0.00	1.00
163.78	49.27	163.78	49.26	0.00	0.00	1.00
175.12	37.40	175.06	37.38	-0.06	-0.02	1.00

Table 2: Mode pair table for highly damped rotor model.

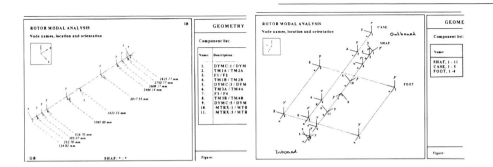

Fig 4: Wireframe model of boilerfeed test pump. Left side : measurement locations and directions along the rotor. Right side : measurement locations and directions along the rotor (shaf), the top of the pump barrel (case) and the top side of the pump pedestals (foot).

Fig 5 : Boilerfeed testpump equipped with AMB

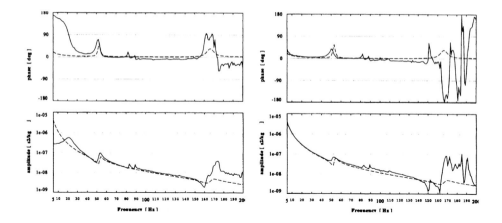

Fig 6a: Experimental FRF (between SHAF:4:-Y and SHAF:3:+Y) based on excitation forces, calculated from currents and positions (solid line). FRF from FE model (dashed line). Dry rotor.

Fig 6b: Experimental FRF based on excitation forces, calculated via strain gauge force transducers and inertia effects (solid line). FRF from FE model (dashed line). Dry non-rotating rotor.

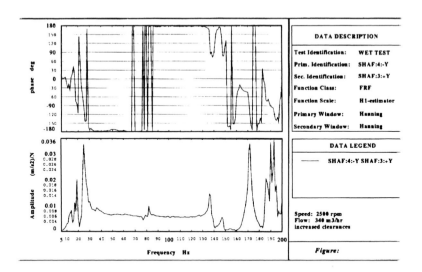

Fig 7: Experimental FRF for pump rotor at 2500 rpm, medium clearances

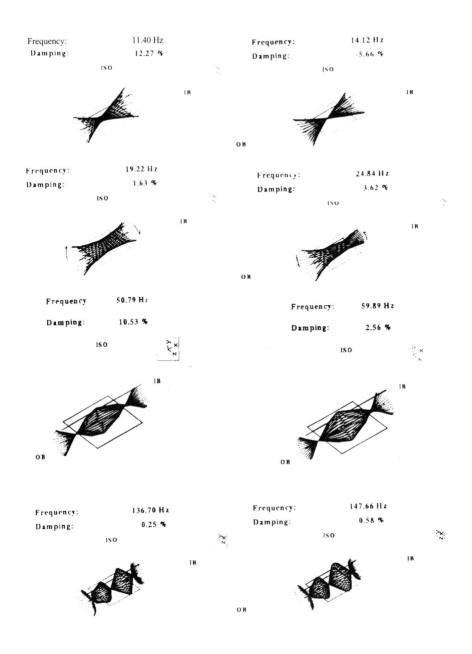

Frequency: 11.40 Hz
Damping: 12.27 %

Frequency: 14.12 Hz
Damping: -5.66 %

Frequency: 19.22 Hz
Damping: 1.63 %

Frequency: 24.84 Hz
Damping: 3.62 %

Frequency   50.79 Hz
Damping: 10.53 %

Frequency: 59.89 Hz
Damping: 2.56 %

Frequency: 136.70 Hz
Damping: 0.25 %

Frequency: 147.66 Hz
Damping: 0.58 %

Fig 8: Modeshapes of the rotor with "medium clearances". Left hand side: BW modes. Right hand side: FW modes.

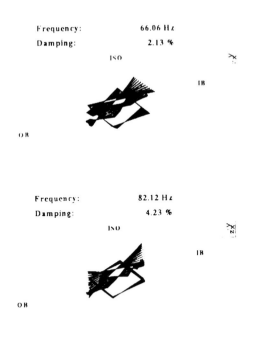

Frequency: 66.06 Hz
Damping: 2.13 %

Frequency: 82.12 Hz
Damping: 4.23 %

Fig 9: Coupled rotor-casing modes

	Mode No	Freq. Hz	Input Shaf: 3 +Y	Input Shaf: 3 +Z	Input Shaf: 7 +Y	Input Shaft: 7 + Z	All
1	1	11.4	70.2	40.8	100.0	67.6	31.3
2	2	14.1	60.8	49.2	100.0	60.9	28.4
3	3	19.2	58.2	100.0	70.7	40.2	14.1
4	4	24.8	100.0	97.1	38.4	40.8	16.5
5	5	50.8	46.3	50.6	77.5	100.0	5.0
6	6	59.9	41.3	57.0	77.2	100.0	2.5
7	7	66.1	14.1	89.6	22.5	100.0	0.5
8	8	76.7	22.5	10.9	100.0	21.5	0.2
9	9	82.1	100.0	32.0	76.6	48.6	0.3
10	10	136.5	40.3	49.8	100.0	83.6	0.8
11	11	147.4	29.4	31.5	98.3	100.0	0.4
12	All		24.8	22.7	30.5	22.0	100.0

Table 3: Sums of modal participation factors, pump with clearances from test 2 (Numbers are listed in table 4).

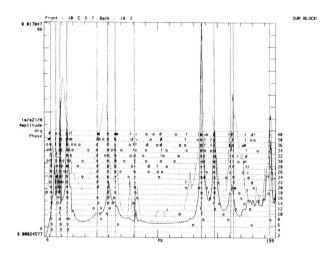

Fig 10: LSCE stabilization diagram and Sumblocks function (Black : amplitude. Grey : phase) for boilerfeed testpump with AMB

	Test 1	Test 2	Test 3
Balance sleeve	483 ± 25	584 ± 25	711 ± 25
Hub wearings	584 ± 25	864 ± 25	1016 ± 25
Eye wearings	787 ± 25	787 ± 25	1016 ± 25

Table 4: Increasing diametrical clearances (microns) used in subsequent modal tests.

	Test 1	Test 2	Test 3
First bending mode, FW	53.7 Hz 15.6%	50.8 Hz 10.5%	50.5 Hz 9.2 %
First bending mode, BW	64.1 Hz 6.6 %	59.9 Hz 2.6%	58.9 Hz 1.4 %
Second bending mode, FW	136.2 Hz 0.2%	136.7 Hz 0.2 %	137.2 Hz 0.7 %
Second bending mode, BW	147.5 Hz 0.5 %	147.6 Hz 0.6 %	147.8 Hz 0.8 %

Table 5: Eigenfrequencies and damping ratios from experimental modal analysis, rotor with increasing clearances.

Fig 11: Field exciter installed around coupling end of boilerfeed pump shaft

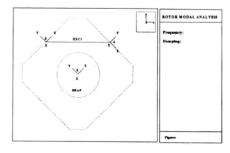

Fig 12: Wire frame model of the field test exciter and of the centre of the shaft.

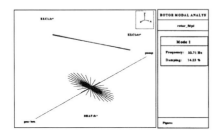
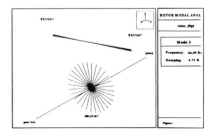

Fig 13: BW (left side) and FW (right side) 'modeshapes' of the boilerfeed pump

C556/005 © IMechE 1999

C556/013/99

Improving the accuracy of reciprocating compressor performance prediction by considering real gas flow in the valve chambers and associated piping

D NINKOVIĆ
R&D Group, Sulzer–Burckhardt Eng. Works Limited, Winterthur, Switzerland

SYNOPSIS

Presented in the paper is the outline of a mathematical model describing unsteady real gas flow in a reciprocating compressor plant. A modified version of the Lax-Wendroff algorithm is used to compute the real gas flow in the pipes, the former being connected to the boundaries by means of a real gas version of the method of characteristics. Pipe boundaries are modelled as generalized throttling elements. The interim results show that accounting for the real gas flow in the compressor attachments increases the accuracy of performance prediction, and helps optimizing the valves under more realistic conditions.

NOTATION

A	cross-section (flow) area,
C_d	discharge coefficient,
h	specific enthalpy,
$\dot{m}$	mass flow rate,
p	pressure,
T	temperature,
u	specific internal energy,
w	gas velocity,
κ	ideal gas isentropic exponent ($= c_p / c_v$),
ρ	gas density,
$( )_0$	total (stagnation) value.

1. INTRODUCTION

From the standpoint of a reciprocating compressor manufacturer, the oil, petrochemical, and general process industry is characterized by ever changing specifications as regards the gas mixture to be compressed, and the operating conditions, i.e. pressure and temperature ranges, involved. Offering an optimal machine in the case of a near-repeat order does not represent a difficulty; it is the new applications, both in terms of gas mixture and the operating conditions, that present a problem in satisfying the accuracy of the performance data stipulated by the customer, while at the same time staying consistent with the budget available and other constraints. Experience and/or accurate performance prediction are needed in order to arrive at an optimal solution in the latter case.

Considering the operating conditions, a trend has been emerging in the recent years to control the delivery rate of the compressor by varying its speed. This trend has been brought about by the availability of relatively reasonably priced variable frequency converters able to drive common asynchronous electric motors in the megawatt power range. This represents an attractive solution to the recip capacity control problem for both the machine manufacturer and the user, as no modifications to the compressor itself are required; and the fact that the delivery rate is nearly linearly dependent upon its speed, makes for an easy integration of the compressor control into the plant control system. However, the variable speed mode of operation presents new problems to the compressor manufacturer on at least three counts, in that the drive chain is more susceptible to torsional vibrations, there is an increased risk of the variable frequency pressure pulsations exciting resonances in the pipework, and optimizing the valves becomes more difficult. Availability of good quality software for performance simulation is of paramount importance in dealing with the latter two problem areas.

In an earlier paper (1), a program was described for performance prediction (i.e. sizing) of reciprocating compressors working with real gas mixtures over a wide range of operating conditions. The underlying model used simplified boundary conditions for the real gas flow through the valves, in that constant pressure was assumed in the suction and delivery valve chambers. In this manner the cylinder was effectively decoupled from the gas dynamics in the neighbouring vessels and piping. Although the resulting software is capable of making fairly accurate performance predictions, its shortcomings in the cases where significant pressure pulsations are present in the valve chambers were also pointed out in (1). But it is not only the accuracy of the performance prediction that is at stake here, for it is an established fact that gas state fluctuations in the valve chambers affect the compressor performance by way of interacting with the valve dynamics (see e.g. (2) and the references cited therein). It is precisely the interaction of these two dynamic phenomena that needs to be addressed in the simulation model if a successful valve optimization is to be realized. These factors prompted further work aimed at extending the scope of the model discussed in (1).

The minimum model extensions needed for taking into account the interactions between the valve and gas dynamics external to the cylinder proper must include the valve chamber (plenum), and the pipe connecting the latter to the relevant pulsation damper. These are the very elements that the valve "sees", as the authors of (2) succinctly put it. However, the pulsation damper must also be accounted for in the model, since it defines the resonant frequency of the gas path between the valve and the damper. Furthermore, the model must be capable of predicting compressible flow of real gas mixtures over the entire range of operating conditions of interest.

A word of clarification is in order at this juncture, for the phenomena mentioned above belong also to the well-known field of pressure pulsation simulation. A pressure pulsation study has become an almost obligatory part of the compressor plant design documentation, the purpose of it being to prove that pressure-induced vibrations in the pipework lie below limits stipulated by national or international standards or recommendations (e.g. API618 (3)). The study is concerned with acoustic phenomena in the pipework, i.e. acoustic wave generation by the pulsating flow pattern of the reciprocating machine, wave propagation and reflection, and the existence of standing waves. It is instrumental in locating possible acoustic resonance conditions and devising measures to detune or otherwise alter the system in order to avoid exciting them. It has traditionally been performed by means of analog or hybrid computers, which have been virtually superseded by fast digital programs.

The foundation of the classical pressure pulsation computation methods is the so-called acoustical approximation, which basically consists of neglecting the mass transfer (i.e. the gas flow) when solving the governing equations of pipe flow (4). Since the waves propagate at the local speed of sound, the latter is also the chief parameter affecting the calculation accuracy. It is an input variable to the program, usually computed externally by means of an e.g. sizing or gas property calculation program. Yet it is the very variable dependent upon the gas flow itself, i.e. upon the pressure and temperature distributions in the pipework, which in turn are interrelated with the velocity of gas flow.

The main difference between the acoustical methods and the approach to be outlined below consists of dispensing with the acoustic approximation in the latter case. The real gas dynamics method thus developed aims at performing potentially more accurate simulations of the cylinder neighbourhood, and not at being used as a pulsation study tool. Furthermore, accounting for the interaction between valve dynamics and the pressure pulsations is now being stipulated by the Design Approach 3 of API618 Ed. 4 (3) also at the acoustical level (indeed, a well-known commercial acoustic pulsation study code has recently been augmented by a valve dynamics model (5)). It was therefore felt that exploring further possibilities for improving the accuracy of the performance prediction and valve optimization by considering the effects of real gas flow in the model was worth the extra effort.

2. DESCRIPTION OF THE MODEL

2.1. Previous work

The field of one-dimensional gas dynamics is characterized by relative scarcity of commercial software for computing compressible flow of gases whose behaviour strongly deviates from the ideal gas law. This statement refers to the field of relatively "cold" (-180 to +250°C), moderate Mach number ($Ma < 0.5$) compressible real gas flow, for there is a well developed area of high temperature, high velocity gas dynamics pertaining to various aspects of jet and rocket propulsion. To the author's knowledge, the recent book by Rist (6) is the only comprehensive work devoted to the subject of the above specified subset of real gas dynamics (referred to as "Thermogasdynamics" in the book), if not necessarily written from the standpoint of CFD (computational fluid dynamics). Most of the work performed so far in this area has been prompted by the interest of companies operating large natural gas distribution networks, which is reflected in the use of specialized (i.e. rather narrow scope, but high accuracy) gas models, and the proprietary character of the codes developed. In addition, practically all of the published algorithms for computing one-dimensional compressible gas flow assume ideal gas behaviour. Even the oldest technique for calculating one-dimensional gas flow, namely the method of characteristics, has only comparatively recently been updated to handle real gases (7).

Turning back to the simulation of gas flow in reciprocating compressor plants, the first application of modern numerical methods to this problem is due to Benson and Üçer (8). Practically at the same time, a noteworthy activity was taking place at the University of Strathclyde (e.g. (9), see also (10) for a review of the state-of-the-art as of 1982). The models described in (8) and (9) were similar in that they both used a second order finite difference scheme to solve for flow variables in the pipework, based on the ideal gas assumption. Unfortunately, the work done in the area of ideal gas compressor flow simulation in those fruitful years did not lead to any publication worth of note from the standpoint of real gas dynamics thereafter, with a possible exception of a hint at the use of real gas parameters in the computation of the pipe gas flow in (11).

Since a reliable cylinder model of (1) was already available at the outset of the project phase outlined in this paper, the approach adopted was to find a program capable of good quality simulation of unsteady pipe flow of ideal gas, extend it in order to cater for the real gas effects, and connect it to the cylinder model. Among several alternatives, a proprietary program used by several major European IC engine manufacturers was chosen as the basis for this development (hereafter referred to as the kernel). In addition to a proven track record in the area of reciprocating machinery, the program is backed by an extensive experimental data base pertaining to various elements commonly found in a pipework, such as various kinds of T-joints, etc.

The kernel differentiates between unsteady gas flow in pipes, and quasi-steady throttling flow at the pipe ends (12). The throttling device at a pipe end represents a generalized resistance element, and can be as simple as an orifice or as complex as an engine cylinder. A complex system can thus be built up of well-defined basic elements in a systematic manner, using always the same interface to connect the pipes with other components. Adding a new element to the system involves modelling the processes taking place internally, and respecting the interface formalism when connecting the former to the pipes.

2.2. The cylinder model

Although the cylinder model is documented in (1), its salient features will be repeated here for the reader's convenience. The model consists of differential equations of mass and energy balance for the gas trapped within the cylinder control volume, closed by a real gas equation of state. The cylinder walls are non-adiabatic; and their temperature is kept constant within a single cycle, but allowed to vary from one cycle to another until steady state is reached. A comprehensive model of the interaction between the valve sealing element dynamics and the real gas flow through the valve passages is also included.

The assumption of constant pressure at the system boundaries stipulated in the original model of (1) has now been dispensed with by modelling the suction and delivery valve plena as constant volume, variable gas state vessels, communicating with the neighbourhood by means of the above mentioned generalized throttle and pipe mechanism. The extensive know-how gained in the process of developing the cylinder model has been fully brought to bear on the valve plenum model. This includes heat transfer between the gas and the plenum walls separating it from the coolant and the outside air (or, indeed, an ice cover in the case of low suction temperatures), time-varying, cycle-constant wall temperatures, and a real gas version of the energy balance equation.

The time-dependent governing equations of the composite cylinder model are integrated numerically by a predictor-corrector algorithm within the iterative computation framework of the kernel.

2.3. The pipe flow model

The non-linear, hyperbolic partial differential equations describing the pipe flow are solved in the kernel by means of the well-known Lax-Wendroff two-step algorithm (e.g. (9)). Since the governing equations in their native form are not restricted to a particular gas model, it is only the computational part of the algorithm that has to be modified in order to make it capable of calculating the real gas flow. Even so, the modifications needed are by no means trivial, since the real gas correlations are customarily expressed in terms of pressure and temperature as the independent variables, whereas the pipe flow conservation laws are formulated in terms of the vector $\left(\rho, \rho \cdot w, \rho\left(u + w^2/2\right)\right)$. A direct connection between the two can only be effected by time-consuming iterative computations.

Provided the solution vector does not fluctuate appreciably over a particular pipe segment, one can linearize the gas data correlations around the mean values of pressure and temperature, and thus save a significant amount of computation time. This is the solution adopted in the current version of the real gas Lax-Wendroff pipe flow algorithm.

The pipe model used allows for gradual cross-section variations and wall friction. A simple means is provided for taking into account heat transfer between the gas and pipe wall in that the latter is assumed to have temporally and spatially constant temperature (user-specified).

2.4. The method of characteristics

The pipe flow algorithm needs additional information at the boundaries, which is usually provided by the method of characteristics (9). A full derivation of a real gas version of the latter is presented in (7); it is based on the existence of two distinct, gas state dependent, isentropic exponents for a real gas, e.g.

$$p \cdot v^{\kappa_{p,v}} = const. \quad \text{and} \quad p \cdot T^{\kappa_{p,T}-1} = const. \tag{1}$$

The equations of characteristics thus obtained have the same form as their ideal gas counterparts; it is the implicit differences in the underlying gas models that produce different results in the ideal and the real gas cases.

2.5. The generalized throttling element

This term refers to an adiabatic flow process through a discrete element (orifice, valve, abrupt cross-section change, etc.), whereby pressure loss is incurred under the conditions of constant enthalpy. The mass flow rate through the element is calculated by introducing the discharge coefficient, the value of which is usually obtained through measurements:

$$C_d = \dot{m}/\dot{m}_{th} \tag{2}$$

The theoretical (ideal) mass flow rate corresponds to frictionless isentropic flow from the upstream stagnation conditions:

$$\dot{m}_{th} = A_2 \rho_2 \sqrt{2(h_{01} - h_2)} \tag{3}$$

where the indices 1 and 2 refer to the upstream and downstream conditions, respectively.

In ideal gas flow, the theoretical mass flow rate is calculated by using the standard textbook isentropic formula:

$$\dot{m}_{th} = A_2 \cdot \sqrt{2(p_0 \rho_0)_1} \sqrt{\frac{\kappa}{\kappa - 1} \left[\left(\frac{p_2}{p_{01}} \right)^{\frac{2}{\kappa}} - \left(\frac{p_2}{p_{01}} \right)^{\frac{\kappa+1}{\kappa}} \right]} \tag{4}$$

For several reasons, deriving a real gas version of this formula capable of achieving an acceptable accuracy over the entire pressure and temperature ranges of interest has been the most difficult part of the project. The main differences to the ideal gas case are the existence of different isentropic exponents, and their varying along the gas state change. The fact that $\kappa_{p,v}$ can also assume the value of 1.0 gives rise to a singularity in the standard ideal gas formula for calculating the stagnation pressure, i.e.

$$\frac{p_0}{p} = \left(1 + \frac{\kappa - 1}{2} Ma^2 \right)^{\frac{\kappa}{\kappa - 1}} \tag{5}$$

which only adds to the above two general points regarding the isentropic exponents. The formula set needed must also include a real gas version of the choking criterion.

The solution found after performing many numerical experiments with several derivations (13) was to use suitably calculated average values of the compressibility factor and the isentropic exponent in the real gas version of Eq. (4). Still more complicated is the calculation of stagnation temperature in the ranges with strongly varying state variable derivatives, such as e.g. ethylene along a compression path typical for the LDPE process (1000-3000 bar). Although this calls for iterative procedures, such extreme cases are handled by multi-step explicit approximation formulae in the current version of the program. On the other hand, the choking criterion for a real gas can be computed in a rather straightforward manner.

3. RESULTS

An interim version of the program is currently being tested for plausibility, accuracy, and computation time. The testing proceeds by comparing the computation results obtained with theoretical solutions (e.g. acoustic resonance frequencies), sizing data and maintenance records of customer's plants, and available test data. A comprehensive measurement project is also being set up, aimed at gathering high quality dynamic data for checking the accuracy of the program and developing possible modelling strategies.

The space remaining in the current paper allows us to present but two sets of results obtained by means of the program. The first case is a single cylinder, double-acting, lubricated machine compressing nitrogen at various speeds and pressure ratios on a test stand. Unfortunately, the tests carried out with this machine in 1982 did not include acquiring valve dynamics and pressure pulsation data. The comparisons to be made must therefore be limited to the global results, such as capacity, gas temperature, and indicated power.

With reference to Table 1 below, prediction results of the model outlined in (1), i.e. constant pressure in the valve plena, no pipework (referred to as RGCNA hereafter), and those of a compressor model with attachments (RGCWA) are compared to measured data. Including the cylinder neighbourhood into the simulation clearly corrects the dependence of the capacity error upon the speed displayed by the RGCNA in this case, and increases the indicated power computed. As to the discharge temperature, both models have a general tendency towards somewhat higher values, which indicates that further refinement of the cylinder and/or valve plena heat transfer models may be required. An additional factor affecting the temperature computed by the RGCNA is a higher cylinder pressure in the discharge phase (due to pressure variation in the valve plenum), giving rise to a further increase in the temperature (see Fig. 1).

Presented in Fig. 1 are the traces of cylinder pressure and valve lift against crank angle computed by the two models at 627 min^{-1}. Actually, all the predictions presented were generated by the same program, i.e. the RGCWA; the two straight lines in the pressure plot, depicting the constant pressure in the valve chambers, are the result of setting the respective volumes to a very high value. Due to mismatch between the valves and the operating conditions, there is a bad valve flutter, making this case a real challenge to simulate accurately. The valve lift traces computed by RGCNA display a tendency of both the suction and the discharge valves towards early closure when operating against a constant plenum pressure (second plot from the top). The interaction between the gas and valve dynamics brought about by the inclusion of the valve plena and the associated piping in RGCWA modifies the flutter frequency of both valves (bottom traces), making them close almost on time, i.e. very near to the respective piston dead centres. This improves the cylinder performance, giving rise to the higher delivery rates in the case of RGCWA (see Table 1).

Table 1 Prediction errors in the ideal gas range (deviations from test data)

	Capacity [%]		Disch. temp. [K]		Ind. power [%]	
Speed [min^{-1}]	RGCNA	RGCWA	RGCNA	RGCWA	RGCNA	RGCWA
627	-7.81	-2.24	+6	+9	-3.74	+1.10
950	-0.33	+2.27	+10	+16	-4.41	+1.67
1500	+9.33	+2.50	+10	+16	-4.08	-0.79

Since the gas flowing through the valve into the discharge plenum causes the pressure in the latter to rise, the instantaneous pressure difference across the discharge valve is diminished, giving rise to a further increase in cylinder pressure (in order to overcome the valve resistance). A similar cause-and-effect mechanism is present in the suction phase; each of them tends to elongate the pressure-volume diagram in the direction of the ordinate and thus make the machine consume more power. This explains the higher values of indicated power in the case of RGCWA in Table 1.

The second example to be presented here deals with the other end of the gas spectrum: the simulation object is the second stage of an ethylene hypercompressor, operating between 1200 (suction) and 2900 bar (delivery). The gas undergoes a variation in the compressibility factor from 2.5 to 4.7 in the cylinder; and the gas density in the delivery valve plenum is about 60% of the water density. The high value of the *p-v* isentropic exponent (9.5 and 7.5 at the suction and delivery conditions, resp.) gives rise to steep pressure vs. volume gradients, calling for short integration steps in the program. The machine possesses four identical second stage cylinders, calculated for a total delivery rate of 97000 kg/hr ethylene at 2900 bar. A single cylinder has been simulated here, consistent with the RGCNA model assumptions.

The upper two plots of Fig. 2 show the traces of cylinder and valve plena pressure and valve lift against crank angle computed by the RGCNA model at the nominal running speed of the machine (200 min^{-1}). Apart from a hint at the discharge valve closing slightly late, they portray a normally functioning compressor stage.

The curves presented in the lower two plots of Fig. 2 refer to the same cylinder, but with the closest attachments added (valve plena and channels, and a length of pipe at each cylinder

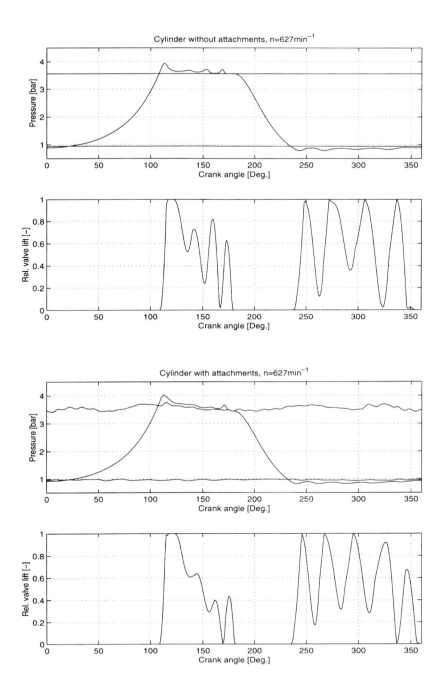

Fig. 1 Pressure and valve lift curves computed by RGCNA and RGCWA (ideal gas)

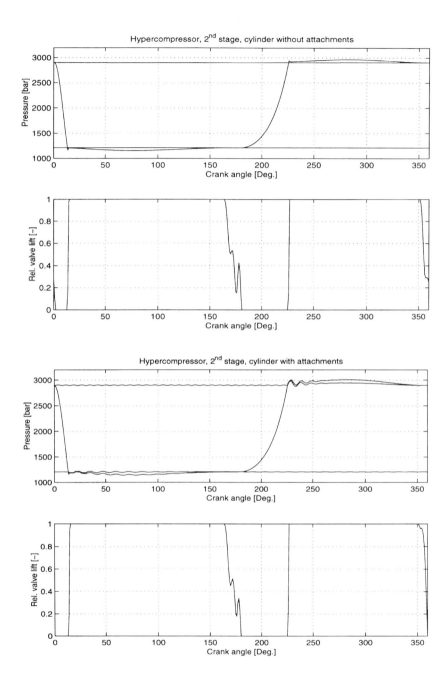

Fig. 2 Pressure and valve lift curves computed by RGCNA and RGCWA (real gas)

side). Interactions between the cylinder and its neighbourhood are quite evident in the pressure traces; however, they do not seem to disturb the valve function. As a matter of fact, the discharge valve improves its function in that it closes at the top dead centre now.

Since no test data are available for this machine, one must resort to comparing the prediction results to each other, and possibly to maintenance reports for similar plants. The former are summarized in Table 2 below.

Apart from the capacity being equal in the two predictions, the discharge temperature and indicated power display the same pattern as in Table 1, i.e. both are somewhat higher in the RGCWA prediction (4 K and 2.24%, resp.). The experience this company has with ethylene hypercompressors suggests the higher end temperature value to be right. Regarding the power figure, no comment can be made unless the whole compressor (all eight cylinders) has been simulated, since only the total power consumed by the machine can be measured.

Table 2 Performance prediction results for an ethylene hypercompressor, 2nd stage

	Capacity [kg/h]		Disch. temp. [°C]		Ind. power [kW]	
Speed [min⁻¹]	RGCNA	RGCWA	RGCNA	RGCWA	RGCNA	RGCWA
200	97700	97700	84	88	8740	8936

A word of caution regarding the latter performance prediction is in order, for although the cylinder attachments are now accounted for in the model, it is still an individual cylinder that has been dealt with in the simulation. A machine of this kind feeds an intricate pipework (normally without a pressure pulsation damper), making complex interactions between the individual cylinders both possible and likely. A simultaneous simulation of the entire stage (i.e. all four cylinders) is thus very desirable from the standpoint of accuracy. On the other hand, extremely short integration step sizes dictated by the fast process dynamics result in a large number of meshes in the pipes, which together with the iterative character of the computations gives rise to unacceptably long computation times (the configuration simulated here, comprising one single-acting cylinder plus two 1.5 metre long pipe segments, takes about ten minutes CPU time on a Pentium Pro 200 MHz to complete a revolution; and many are needed before steady state is reached). Such a simulation is therefore yet to be performed by this program version, although advances in the computer technology and improvements in the algorithms may make this goal viable sooner than one may endeavour to think.

4. CONCLUSIONS

A novel reciprocating compressor performance prediction model has been presented, capable of simulating not only the cylinder proper, but also unsteady flow of real gas in the valve chambers and the associated piping. The interim results obtained so far suggest that the prediction accuracy has benefited from the model extensions, in that the tendency of an earlier model that assumed constant pressure at the cylinder boundaries towards too optimistic results (higher capacity and lower indicated power) has been corrected. Interaction between the valve function and real gas dynamics in the adjacent compressor elements has been demonstrated, providing new possibilities for valve optimization under more realistic conditions.

Work is now proceeding towards gathering high quality dynamic data on the test stand, aimed at perfecting and/or calibrating the valve dynamics and heat transfer models used.

5. ACKNOWLEDGEMENTS

The author is grateful to the management of Sulzer-Burckhardt Eng. Works Ltd. for the opportunity to realize this project, and the permission to publish the results obtained. This project is a result of a teamwork with Dr. H.-J. Linnhoff and Dr. P. Schindler, both of Ing.-Büro Linnhoff, Bochum, Germany, whose programming contributions and many stimulating remarks regarding the theoretical derivations are hereby thankfully acknowledged.

6. REFERENCES

(1) Ninković, D. Performance Prediction of Reciprocating Compressors Working With Real Gas Mixtures Over a Wide Range of Operating Conditions. Proc. Sixth European Congress on Fluid Machinery for the Oil, Petrochemical, and Related Industries, The Hague, 1996, pp. 169-177.

(2) MacLAREN, J.F.T., TRAMSCHEK, A.B. Prediction of Valve Behaviour With Pulsating Flow in Reciprocating Compressors. Proc. 1972 Purdue Comp. Tech. Conf., pp. 203-211.

(3) Reciprocating Compressors for Petroleum, Chemical, and Gas Industry Services. API Standard 618, 4th Edition, June 1995.

(4) SINGH, R., SOEDEL, W. A Review of Compressor Lines Pulsation Analysis and Muffler Design Research (Part II). Proc. 1974 Purdue Comp. Tech. Conf., pp. 112-120.

(5) Optimising Valve Configurations. PULSIM Newsletter, TNO Institute of Applied Physics, 2600 AD Delft, The Netherlands, Jan. 1998.

(6) RIST, D. Real Gas Dynamics (in German, or. "Dynamik realer Gase"). Springer, 1996.

(7) FLATT, R. On the Application of the Numerical Methods of Fluid Mechanics to the Dynamics of Real Gases (in German, or. "Zur Anwendung numerischer Verfahren der Strömungslehre in der Realgasdynamik"). Forschung Ing.-Wesen, 1985 (51), pp. 41-51.

(8) BENSON, R.S. and Üçer, A.Ş. A Theoretical and Experimental Investigation of a Gas Dynamic Model for a Single Stage Reciprocating Compressor With Intake and Delivery Pipe Systems. J. Mech. Eng. Sci. 1972 (14), No. 4, pp. 264-279.

(9) MacLAREN, J.F.T. et al. A Comparison of Numerical Solutions of the Unsteady Flow Equations Applied to Reciprocating Compressor Systems. J. Mech. Eng. Sci. 1975 (17), No. 5, pp. 271-279.

(10) MacLAREN, J.F.T. The Influence of Computers on Compressor Technology. Proc. 1982 Purdue Comp. Tech. Conf., pp. 1-12.

(11) PEREVOZCHIKOV, M.M. and CHRUSTALYOV, B.S. Theoretical and Experimental Researches of Unsteady Gas Flow in the Pipeline of the Reciprocating Compressor. Proc. 1994 Purdue Int. Comp. Eng. Conf., pp. 515-520.

(12) GÖRG, K.A. Computation of Unsteady Flow in the Pipework of IC Engines Under Special Consideration of Multiple T-Junctions (in German, or. "Berechnung instationärer Strömungsvorgänge in Rohrleitungen an Verbrennungsmotoren unter besonderer Berücksichtigung der Mehrfachverzweigung"). Ph.D. Thesis, Ruhr-Univ., Bochum, 1982.

(13) NINKOVIC, D. Calculating the real gas isentropic state change. 1994, Unpublished work for internal use, Sulzer-Burckhardt Eng. Works, Winterthur, Switzerland.

C556/003/99

Retrofit of a high-speed synchronous motor

O VON BERTELE
Consultant, Norton, UK

A high speed motor was retrofitted to the compressor train of a nitric acid plant in 1984. This paper describes the reason for the choice of such a motor, the problems that had to be overcome and the experience of operating the unit for 14 years. The unit has recently been decommissioned but will probably be rebuilt in China.

1 BACKGROUND

Nitric acid plants with power recovery are basically very inefficient gas turbines. The fuel is ammonia and the aim is to maximise the production of nitrous gases in the combustor. After these gases have been removed in an absorption tower to make acid the tailgases are reheated and expanded. The power output from the expander, being insufficient to drive the air/gas compressor, has to be supplemented by an electric motor or a steam turbine.

Thirty (30) years ago self contained plants were popular. One nitric acid plant built in Belfast in the early 70ties used the steam raised by the combustion of ammonia to drive the combined air and nitrous gas compressor with a condensing steam turbine, in addition to the expander. Fig 1 is a flowsheet of the plant after it was modified.

On the same site there were also some oil fired boilers raising steam for general duties. When the oil crisis came the question was raised why condense steam and then raise low pressure steam for mundane drying duties, why not stop condensing and use the steam raised in the nitric acid plant for these duties?

Lifting the backpressure to 2 bar and not condensing the steam in the compressor driver, would have reduced the power output of the turbine by about 2MW. How could one replace this? Various schemes were considered. One was to remove the turbine altogether and drive the compressor with an electric motor through a gearbox and install a backpressure turbine, driving an alternator, to efficiently let the generated steam down. This however was not possible for a number of reasons:

1) The electricity supply to the site was only 5MVA with a corresponding low fault level. This was not sufficient to run the motor, in particular a much larger supply was needed for start-up. Fig 2 shows the power required by the air and nitrous gas compressor against speed as well as the power generated in the expander. Note that a direct on line start requires a motor power of over 8 MW. (throttling the compressor so that its suction is at a vacuum lowers the power demand somewhat) Raising the speed to 75% (5,500 rpm), the speed at which the burner can be lit, however needs a motor power of only 3 MW.

2) There was insufficient space in the compressor house for a motor and gearbox.

3) The length of a shut down required to carry out such a modification could not be tolerated

4) The existing oil unit was too small to allow a gear box to be supplied with oil. A new oil system would therefore be required.

5) The plant is built on reclaimed land in Belfast dock and the machine foundation was supported by 30 m long piles. Time and space constraint made it impossible to fit new piles to take the additional load of a normal motor and gearbox.
Another scheme investigated was to use a variable speed motor driving through a gearbox.

This scheme, as already stated above, would have allowed the use of a smaller motor as the unit could have been started by bringing the speed up gradually. Full speed could then be reached once the burners had been lit, and power became available from the expander. However this scheme too suffered from the major foundation modifications needed. It was therefore also discarded.

The most promising solution appeared to be replacing the condensing turbine with a passout one and fitting a high speed motor to drive through the turbine to make up the lost power. This scheme was adopted.

2 PROBLEMS

Once the decision was made to use a direct drive there were two major problems: lack of space and lack of time.

The high pressure steam lines of the nitric acid plant were only one meter away from the end of the condensing turbine. Rerouting them was not an option. Further, fertiliser demands required any modification to be carried out in less than four weeks.

3 SOLUTION

3.1 Foundation

The whole plant had been built on reclaimed land in Belfast Loch. The original foundation was built on piles 30m long. It was not feasible to install more piles with the unit running. It was therefore decided to attach a bracket to the existing 6m high, foundation to carry the motor. The bracket was supported by anchored feet standing on the old foundation and was tied to the top table with epoxy set rawlplugs. (Fig 3)

The highest first critical of the foundation (vertical movement) was estimated to be about 25Hz, well below the operating speed of the unit, which ranges from 90 to 125Hz. No attempts were made to calculate higher modes, as avoiding all criticals in the operating range was not feasible. Care however was taken to maximise damping in the extention of the foundation.

Additional loads to the dead weight of the foundation and the compreeorset will be caused in case of a short-circuit in the motor. However even the most severe short circuit was found to alter the reaction forces in the table legs of the foundation by less than +/-10%. This was acceptable.

3.2 Electrical Drive .

The variable speed drive had three main parts:
> The motor
> The rectifier/invertor
> A step down transformer.

3.2.1 Motor

The motor is a synchronous machine. Therefore it can supply reactive power to the invertor which in turn commutates the motor. At the time when the motor was ordered 3.5 MW at 7,500 r/min was close to the limit of past experience. Today 40 MW motors running at 6,000 r/min are available as a standard design.

The motor details are as follows:

Rated power	3.5	MW
Operating speed range	5,150 to 7,720 (also 300-5,150 short time)	rpm
Overspeed test	9,300	rpm
Critical speeds	1st 3,920 2nd 9,880	rpm
Rated Voltage	2.2	kV
Noise level	85	dbA at 1m
Total weight	7,350	kg
Rotor weight	1,000	kg
Bearings:	tilting pads.	
Cooling	CACW	

Full load torque could be used over the whole operating range. Continuous operation however, at speeds below 5,150 rpm, is not possible, as it would lead to overheating of the motor due to the low speed of the shaft mounted fans. This is not a problem as extended running at low speed is never required.

The cooler was fitted with double skin tubes so that any leakage could be detected and an alarm sounded.

It was not possible to achieve the required noise level of 85db at 1m with out a noise hood. A hood was therefore fitted to cover the motor and its coolers. The slipring unit however was outside the hood to allow easy access to the brushes needed to be changed three times per annum.

3.2.2 Excitation
Excitation was via sliprings. This was for three reasons:

1)   No experience was available with diodes rotating at speeds up to 7,720rpm

2)   Brushless excitation would increase the axial length of the motor beyond the available space. Further it would not be possible to mount such a device overhanging the free end of the motor. This would have resulted in the second critical shaft speed to be lowered to about 6,300rpm, i.e. in the middle of the operating range. There was neither sufficient space for a third bearing to support the free end of an exciter nor was such a solution deemed satisfactory.

3)   Nitric acid plants need a catalyst change every three months which requires the plant to be shut down for two or three days. Provided brush life is at least 4000h the need to change brushes does not cause any significant cost or inconvenience.

3.2.3   Rectifier and Invertor.
The rectifier part of the converter with its smoothing choke acts as a variable direct current generator with an output current adjusted to the required motor torque.

Of particular concern was the wave form distortion caused by rectifiers and invertors which can be reinjected into the mains. As stated above the feeder to the site had a very low rating and the proposed motor would account for 50 % of its capacity, all other drives on the site being small induction motors. The harmonic problem was minimised by the choice of the convertor system and solved by the addition of a filter to the 6.6kV supply system. The filter was combined with a power factor capacitor which helped to improve the power factor close to unity at a load of 2.5 MW at full speed.

The design conformed to the requirements of the electricity board (G5/3)

To minimise harmonics a 12 pulse rectifier bridge and a transformer with two secondary windings, phased at 30° to each other, was selected. In the rectifier two thrysistors in series were used in each arm. Because of the higher frequency of the invertor there were three thrysistors in each arm. Altogether there were 60 thrysistors. Cooling was by a closed air circuit, water cooled. This was chosen because of the corrosive atmosphere. The unit was installed in a concrete substation at the factory and shipped completely assembled.

3.2.4 Transformer
The supply to the site was 6.6kV. The motor operated at 2.2kV, the most economic Voltage for this application. To allow for losses in the rectifier/invertor the secondary of the transformer is at 2.35 kV

3.3 Coupling

To connect the motor with the turbine raised the following problems:

1) The degree of misalignment which could be tolerated by the coupling had to be greater than normal due to the two machines standing on very different foundations: the turbine on the original concrete frame and the motor on the built on steel cantilever extension. A parallel misalignment of 1mm was specified.

2) Due to the space limitation, and also the capacity of the oil supply, the motor was built without a thrust bearing. A coupling without end float was therefore called for.

3) The coupling had to transmit 4.5 kNm under normal operating conditions. Further it had to be flexible in the torsional mode as during a short circuit moments seven times larger than full load torque could be generated by the motor. These had to be prevented from being transmitted to the turbine.

A solution was found in a torsion bar coupling 500mm long and 48mm in diameter. (Fig4) Because of the space restriction it was necessary to forge flanges on both the motor and turbine shaft.

Adding the motor to the system added a new torsional critical at 12 Hz. This meant that any oscillations caused by a short circuit while the motor is on full load, which could be at frequencies between 90 and 128 Hz, would be adequately isolated from the rest of the set. The existing criticals were only slightly changed by the addition of the extra inertia and torsion spring. A short circuit during runup is highly unlikely - operation between 6 and 12 Hz would last less than 10 seconds during a star up. Further the motor load in this speed range is less than 20% of full load torque and current and therefore the oscillating torque generated during a short circuit would also be lowered by a factor of five.

3.4 Control

There were four parameters which had to be controlled:

	the speed of the unit	
	the boiler pressure	(25 barg)
	the passout pressure	(2 barg)
and	the factory main	(20 barg)

It was decided to install a dedicated mini computer, not only to control the four parameters in isolation, but also to manage the system during upsets. For example increased HP steam demand would be accomandated by increasing the power provided by the motor; during an electric upset maintaining the speed would have prirority over steam export.

The speed was measured with six magnetic probes facing a 60 tooth gearwheel integral with the turbine shaft. One probe was used for local indication, one for speed control and one was a spare. One manufacturer has since adopted the govenor system designed for this plant as his standard. The other three probes were used in a voting overspeed trip actuator. Note that a conventional mechanical overspeed trip was not practical for two reasons:

1) To fit a mechanical trip would have required the turbine shaft to be lengthened This was not possible.

2)   The motor was coupled to the turbine and the motor torque had to be transmitted through the turbine shaft. Drilling a hole through the shaft to accommodate a conventional overspeed bolt would have weakened it. The alternative would have been to increase the diameter of the shaft near the overspeed trip but this would have increased the axial length even more.

The turbine itself had two controls: Three cam operated inlet nozzle valves on the live steam and a throttle valve admitting 2 barg steam to the condensing section. The cam of the nozzle valve and the throttle valve were operated with conventional oil operated power cylinders of the turbine maker. To convert the electric output from the computer into hydraulic signals devices first designed for the German Leopard tank were used.

Speed, boiler and pass out pressure signals as well as a the set points for these parameters were sent to the computer. A simplified line diagram of the controller is shown in Fig 5. Output signals went to the rectifier and various let-down valves. An additional complication was that the unit had to be able too operate without the motor. The controller was programmed to do this.

3.5 Timing
The conversion had to be completed in 28 days. This was achieved by careful planning. The floor at the free end of the turbine was removed while the plant was still operating. The foundation extention was then fitted. Also an additional oil pump was installed to replace the shaft driven one. This new pump was commissioned during a routine catalyst change; at the same time the original, shaft driven pump, was removed

 Once the shutdown commenced the condensing turbine was removed and its bed plate cut out so that the new bedplate could be fitted. After the new turbine had been aligned to the existing compressor the motor was provisionally aligned to the turbine. The motor was then removed and the top of the extension was filled with concrete. Once this was done the motor was replaced and aligned to the turbine. The unit was started up 28 days after the shut down was started.

4 PERFORMANCE

Of the 18t/h of steam raised by the plant only 15 were required once the motor was installed, 3 t/h were exported to the factory HP main. 10 t/h at 2bar were also exported. Note that a minimum of 5 t/h was needed to prevent the condensing section of the turbine from overheating.

The controller was found to control the speed of the set to +/- one rev at 7,000 r/min both with the motor and with turbine drive only.

The overall efficiency of the electrical drive system was nearly constant over the whole operating range (5,400 to 7,720r/min) and was approximately 94%

The overall power factor on the network was better than 0.95 under all operating conditions.

The gearwheel on the turbine shaft was situated at a node of the first few torsional criticals. Analysing the signal from one of the probes allowed a check for torsional vibrations to be carried out. None were found.

Besides the planned fuel oil savings the output of the plant was 20% higher than previously. A further bonus was the ability to reduce electricity consumption during high demand periods by reducing the speed or by using more steam.

The vibrations on the bearings of the motor were less than 2mm/sec when the unit was commissioned. After a few years of running the motor vibration level increased rapidly. One of the end shrouds of the motor had cracked. While the repair was carried out the unit was successfully run without any passout, reducing the output of the plant by 20%. Once the rotor was repaired and reinstalled the vibrations dropped to their original level.

5 COMMISSIONING

Two problems occurred. The coupling between steam turbine and air compressor failed due to inadequate design. No consequential damage was suffered.
One of the brushes on the exiter overheated. The cause was a bad bond between the graphite and the copper tail.

6 COST

1984 when the motor was purchased a high speed drive was not only the best technical solution but also the cheapest one for this application.. Table one gives a comparison between the cost of installing a conventional motor together with a turbo generator and the high speed unit. Note that the lack of a high powered feeder to the plant significantly influenced the comparison.

Table 1

	Fixed speed Geared	Variable speed Geared	Variable speed Direct
Motor	10	8	15
Gearbox	10	7	-----
Convertor		35	35
Feeder	15	---	-----
Oil unit	5	5	-----
Plant modifications	20	10	-----
Turbine	30	15	20
Alternator	10	10	-----
	100	90	70

7 CONCLUSION

In spite of the novelty of the modification the installation of the new turbine and the high speed motor was completed to plan. No difficulties were experienced in converting and operating the plant. Operation became simpler, reliability was outstanding and the envisaged savings were achieved. The small dimensions and weight of the motor, not usually critical on land applications, allowed the modification to be carried out in the restricted space available. The precautions taken to limit harmonics entering the grid were successful.

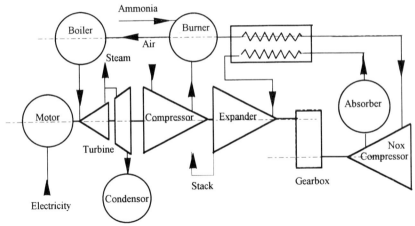

Fig 1 Flowsheet (modified plant)

C556/003 © IMechE 1999

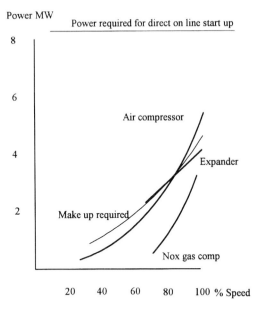

Fig 2 Power requirements

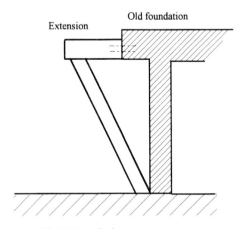

Fig 3 Foundation   Extention

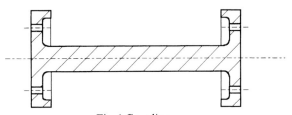

Fig 4 Coupling

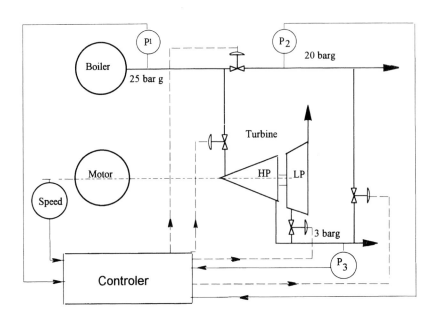

Fig 5 Control system

A solution for deepwater multiphase boosting

V MEZZEDIMI, P RANIERI, and **G FERRARI AGGRADI**
Nuovo Pignone, Firenze, Italy
G DE GHETTO and **M GRANATO**
ENI Agip E&P Division, San Donato Milanese, Italy
A RADICIONI, F RIGHI, and **G D'ALOISIO**
Sasp Offshore Engineering, Fano, Italy

[handwritten signatures]

Further to the successful completion of the long term underwater operations at the ENI-Agip Prezioso Field, Nuovo Pignone, ENI-Agip E&P Division and Saipem Group have undertaken a new R&D project on multiphase transportation and boosting technologies.

A qualification program for a high-flowrate multiphase boosting unit has been foreseen, both considering the aspects of the boosting system and of the subsea unit packaging for deepwater application. The subsea pump is based on the twin counter rotating screws concept, developed by Nuovo Pignone and tested in Prezioso Field.

The design, manufacturing and qualification of the subsea boosting system (flowrate up to 1000 m³/h and pressure rise up to 70 bar) are foreseen, together with the qualification of equipment identified as critical or innovative, which require an industrial qualification.

The paper describes first the operation at Prezioso field, then the new project objectives and scopes, outlines the current status and focuses attention on the subsea unit packaging and on the pump qualification programs. The results of the on-going testing activities carried out (qualification of the pump's seals and of the twin-screw pump capability of withstanding 100% gas volume fraction (GVF) without damage) before performing the full qualification tests of the boosting pump are also presented.

1. INTRODUCTION

The present trend in the offshore oil industry is the development of subsea fields located in deepwater. The exploitation of such a type of fields requires the development of innovative and cost effective technologies one of which is the application of multiphase pumps to boost the untreated well streams. Multiphase production and transportation systems are getting more and more attractive for deepwater applications because they enable the development of marginal fields. In fact they allow:

- to minimise the facilities located in deepwater;
- to locate production facilities far from well locations (e.g. in shallower waters);
- to increase the field productivity.

ENI-Agip and Saipem-Sasp have been active since the mid-eighties developing multiphase boosting systems based on the Nuovo Pignone twin screw pump technology.

2. PREZIOSO FIELD OPERATIONS

The industrial qualification of any developed multiphase boosting equipment requires field operational experience to prove the suitability of the proposed systems in handling untreated well fluids under real operating conditions. The underwater operation poses additional requirements to be fulfilled in order to achieve the expected performance and reliability levels.

To this purpose, after the development of the marine concept of a multiphase twin screw pump, a long term subsea testing campaign was performed.

The characterisation of the hydraulic and mechanical performance of both the pump and its associated auxiliaries constituted the main objective of the tests, together with the assessment of the overall system reliability. This campaign has provided a significant contribution to a better understanding of the performance of such systems in different operating conditions as well as of the issues related to installation, start up and maintenance.

2.1. Marinised screw pump

The pumping system tested included a subsea low size multiphase screw pump (NPV70) in vertical configuration, driven by a water filled electric motor and equipped with a simplified lube oil system; all the system was totally pressure balanced at the process pump inlet conditions by means of special pressure transfer vessels (Fig. 1).

The developed concept of the subsea screw pump resulted in a system configuration that achieves simplicity, integrated unit, insensitivity to water depth, high reliability due to its simplicity, self-regulation without any control requirement.

Improvements have been made in the design of the pump components as well as in the fabrication methodologies and in the selection of the mechanical seals and hydrodynamic bearings configuration. Mechanical seals and bearings are continuously lubricated by a dedicated lube oil system. This system presents a closed circuit connected to an oil tank which is pressure balanced at the pump inlet conditions. A lubrication pump is installed within a pressurised lube oil vessel and is mechanically driven by the pump shaft. This pump provides the required differential pressure to enable the oil circulation and the oil supply to both the mechanical seals and the bearings. Such a differential pressure, which is the maximum load expected across the seals, is practically constant (in a range of 1.5~2 bar) and is totally independent from the pump inlet pressure. The pressure balancing of the whole system is firstly obtained through a "primary transfer vessel" where an elastomeric diaphragm of special shape and formulation separates the lube oil section from the process fluids at the pump suction side. In such a way the lube oil section is continuously maintained at the process pressure without contamination of the fluids. Due to the positive and constant overpressure across the pump seals, the leakage of lube oil will occur towards both the process and the coupling sides, but no leakage of other fluids can occur in the reverse direction. The oil leakage is compensated by the translation of the primary diaphragm and, consequently, by a net volume reduction of the oil reservoir that is maintained at a constant pressure. The oil reservoir can be sized to allow the oil leakage over a specified period of operation according to a pre-defined refilling frequency. The lube oil section also transmits the process pressure to the remaining sections of the system (motor and coupling chamber) by means of two "secondary transfer vessels", which are standard bladder type compensators, that do not pose any particular operational problem in dealing with simple liquids.

The electric motor is also hydraulically isolated from the sea water, being pressure balanced at the pump process inlet conditions through a secondary transfer vessel, that separates the motor coooing water from the primary circuit lube oil. In addition to that, an auxiliary internal impeller provides during operation an overpressure of 0.5 bar to the motor shaft seal chamber to ensure that only water leakage from the motor to the coupling chamber can occur. The transfer vessel also provides for the water storage function; any water leakage towards the coupling zone is compensated by a reduction of the water reservoir net volume, similarly to the primary oil circuit, until an external water refilling is required. The water and oil leakages towards the coupling chamber are collected in another secondary transfer vessel which is pressure compensated similarly to the water vessel. This leakage vessel requires an emptying operation in conjunction with the refilling of the other two vessels.

PRESSURE TRANSFER VESSELS
A: LUBE OIL vs PROCESS
B: LUBE OLI vs WATER
C: LUBE OIL vs LEAKAGE

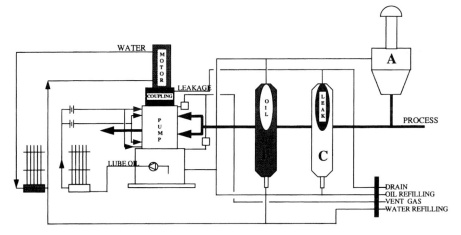

Fig.1 : Schematic of the Prezioso Twin Screw Pump Marine Concept

2.2. Test site characteristics

The adopted configuration for the subsea trials was a "dummy well" like layout with the subsea boosting unit installed in the proximity of the Agip Prezioso Platform.

This oil production platform is located 15 Km south offshore Sicily in 50 m of water and is connected by a 30 km long, 12" sealine to the onshore Oil Centre in Gela. The boosting unit was linked to the platform by two (pump in/out) flexible lines and by an electro-hydraulic umbilical.This solution required particular conditioning and control of the well flow to feed the pump without interfering with the production and allowing suitable bypass of the prototype unit in case of failure. Main pump test parameters (range values) allowed by the surface conditioning facilities have been:

suction pressure	4 to 50 bar;
differential pressure	up to 40 bar
gas void fraction	30 to 90 % (average figures)
oil viscosity	30 to 2000 cSt

Combinations of these parameter values have allowed a comprehensive testing campaign to explore the operating performance of the whole pumping system, in addition to the effectiveness of the pump marine concept.

2.3. Endurance Tests

An endurance testing program for the verification of the overall pumping system long term performance and reliability has been performed; the scheduled 7500 hours of operation have been accumulated. The endurance tests, carried out with the boosting system connected in line to the platform production manifold at fixed operating conditions, showed a high stability with no uncontrolled variations in the wells production. They have been composed of a series of long term testing campaigns with combinations of the operating parameters, that have covered different application scenarios and have investigated specific areas of concern. The main aspects that have been verified and analysed on a long term basis are:

- boosting at high gas void fraction and high oil viscosity;
- pump control by means of variable motor speed or by means of throttling valves;
- interaction of the pumping system with both the wells and the multiphase export line;
- variation of the lube oil pressure across seals and bearings;
- evaluation of any degradation effect on the pump flow capacity over time; this effect has been evaluated also after system recovery.

2.4. Multiphase Boosting System Recovery

When the Module was recovered, the pumping system (pump, motor and auxiliaries) was brought to Nuovo Pignone's plant in Florence, where it was disassembled, inspected, reassembled and tested on the multiphase loop, in the same conditions as the Factory Acceptance Test (FAT) to evaluate possible decay in boosting system performance.

The main results of the check and performance test were:

- all the instrumentation was still in perfect working condition
- the Subsea Telemetry System was still in perfect working condition
- the electric motor filling fluid (water and glycol) was undeteriorated; leakage was minimal, and the related transfer vessel was still full
- the insulation of the motor windings and the high voltage connector was more than 5GΩ; during the new tests (no-load and full-load) electrical input was the same as during FAT.
- all the internal pump components were in very good condition; only the mechanical seals had some signs of wear. Some problem in the seal system had been detected during endurance tests in Prezioso, with increased and discontinuous oil leakage, but it did not prevent the endurance tests from being completed.

2.5. Performance test results.

During the endurance tests in Prezioso the subsea multiphase pump did not have any performance degradation due to erosion, corrosion or wear. During the inspection after disassembling (inspection of the surface conditions both of the screws and of the liner, and check of the gaps) no significant variation , compared to the initial condition, has been found This has been confirmed by the performance tests, that have been conducted using for the liquid phase a mineral oil, ISO VG10, with a viscosity very close to the viscosity of the diesel oil used during the FAT.

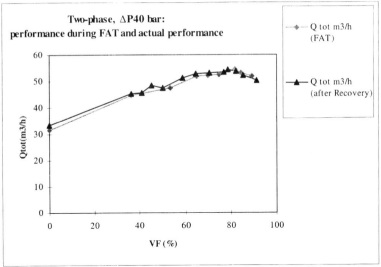

Fig.2: Comparison between pump performance before and after subsea operations.

3. FURTHER DEVELOPMENT

Further to the successful completion of the long term underwater operations Nuovo Pignone, ENI-Agip E&P Division and Saipem Group have undertaken a new R&D project on multiphase transportation and boosting technologies.

Within this project a qualification program for a high-flowrate and differential pressure multiphase boosting unit has been foreseen, based again on the twin screw pump concept tested in Prezioso Field.

The project is named "Deepwater multiphase production and transportation system".

The objective of this project is the characterisation, development and qualification of a boosting unit suitable for handling unprocessed multiphase streams.

The engineering activities of the project have been organised in a conceptual design phase followed by a basic design phase; the construction and the execution of industrial qualification tests of the pump unit prototype and of the whole electrical chain- namely the variable speed drive (inverter), the resistors to simulate the umbilical, the electric motor - as well as the trials on the mechanical seals for the pump, complete the activities foreseen for this project.

This paper mainly deals with the "Deepwater Multiphase Boosting System" (DMBS); the basic philosophy for the proposed DMBS concept is to maximise simplicity and reliability of the whole system. In particular, taking into account the 1000 m water deep operative scenario and the expensive offshore operations in the event of retrieval (for maintenance purposes or main equipment failure) and re-installation, the system reliability has been emphasized. Moreover, no redundancy has been foreseen for the equipment present in the plant, and that operated in Prezioso without problems, in order to minimise weight.

The DMBS subsea system, characterised by a modular configuration, is mainly composed by the base frame and the boosting module.

The former is basically a structure which supports the DMBS module and houses the equipment for tie-ins of umbilical and flowlines, the flowline isolation valves and the inboard

hub of the process connectors; the latter can be subdivided into the following main subsystems:
- pump unit
- electric motor
- process system
- auxiliary system
- subsea control system
- electrical system
- instrumentation
- connection systems

The conceptual phase of the project has highlighted some equipment that are identified as critical or innovative, and that for this reason require an industrial qualification test programme.

In particular, as far as the boosting unit is concerned, the equipment requiring further analysis are the following:
- Pumping unit
- Pump mechanical seals

3.1.Pumping unit

The pumping unit NPV1600 (fig.3), used in the DMBS, is based on the twin screw technology; it has a vertical configuration, with the oil tank acting as the baseplate for the pump and motor stack-up, arranged in the lower part. Its weight and size have been optimised to be as low as possible and its components are designed in order to be easily housed within a subsea module.

The pump is driven by the electric motor by means of a mechanical coupling installed on top of the pump. The mechanical coupling is contained in a suitable chamber filled with the electric motor cooling fluid. The multiphase pump lubrication circuit is composed by the oil tank, where the lube oil is stored, and by the lubrication pump installed inside the oil tank and driven directly by one of the screw shafts. The lube oil is used in the journal bearings and also in the seals to flush and cool them. The lube oil pressure is compensated with respect to the inlet process pressure, in order to maintain the differential pressure across seals to a limited, constant value of about 2-3 bar. The pump is able to guarantee the required performance also when handling solids of dimensions up to 0.25 mm. The maximum sand concentration for continuous operation has been estabilished at 500 p.p.m.

NPV1600 design parameters are:
- max flowrate   800 m3/h at 60 bar differential pressure
- max differenzial pressure 70 bar
- max absorbed power   2500 kW

3.2. Full qualification tests of the pumping unit

The pump will be qualified in a new multiphase loop now under construction in Florence, that will allow multiphase pumps to be tested with flow rate up to 1000 m3/h, suction pressure up to 30 bar, differential pressure up to 70 bar and GVF from 0 to 100%; the loop will be able to create gas and liquid slugs and to simulate the subsea operating condition: a constant delivery pressure irrispective of the pump flow rate.

The new multiphase loop will be completed by december '98.

In order to select for the new loop the optimum solution to test the DMBS pumping unit in all possible subsea operating conditions and to perform some preliminary tests, a medium size multiphase pump, NPV280, vertical configuration, has been assembled in the old multiphase loop in order to verify:

- the maximum GFV at which the pump can work properly
- the possibility to run at 100% gas, and for how long
- which parameters have to be measured, to understand that the pump is running at 100% gas
- the performance of the pump in this condition
- the best way to restart after a gas locking, if any

The pump has been instrumented with position probes to control the gap between liner and shafts, and with thermocouples to control the liner temperature close to the gap (fig.4).

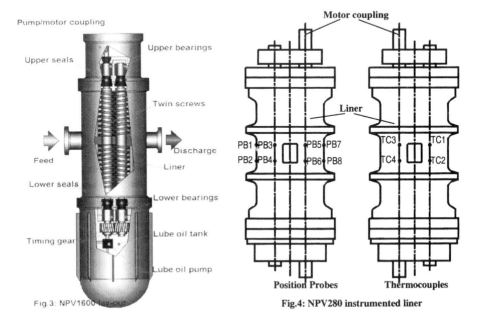

Fig.3: NPV1600 layout

Fig.4: NPV280 instrumented liner

Moreover, the liner has been drilled in the middle, at one pitch from delivery, and the holes have been connected by a piping to an external delivery separator, in order to verify the efficiency of automatic liquid injection in the delivery zone, in the event of gas locking.

The tests performed, that will be repeated on the DMBS pump, have proved that:

- a twin screw pump can run with 100% gas indefinitely (more than six hours test), if the differential pressure is zero and the absorbed power is minimum (only mechanical losses)
- a twin screw pump can give a differential pressure even with 100% gas; the higher the differential pressure the lower the volumetric efficiency, and the higher the internal recycle
- the absorbed power is independent (apart from viscosity losses) from the GVF and is only dependent on the flow rate (i.e. rotational speed) and differential pressure
- so, in the long run at 100% GVF, there is an increase in the internal temperature of the pump (shafts and liner) and a decrease in the nominal gaps
- to avoid contact between the two shafts or between the liner and the shafts the internal temperature must remain within the design limit, but critical temperatures are reached only after several minutes, and safety automatisms can easily be actuated (cooled recycle or liquid injection)

- injection in the pump suction of 3-4 % of liquid is sufficient to reduce temperature to normal operating values.

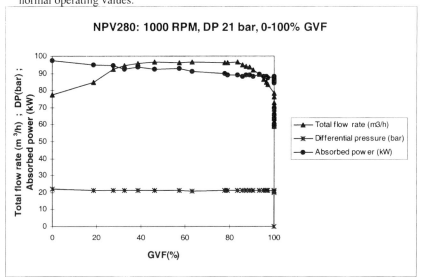

Fig.5: NPV280- Test at 1000 RPM and 21 bar differential pressure.

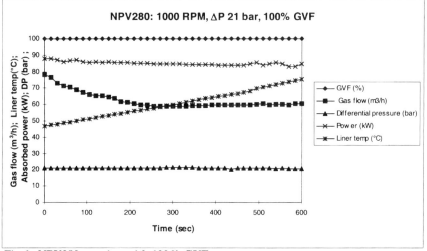

Fig.6.: NPV280 running with 100% GVF

Testing for the present has been performed at low speed (1000 RPM) and low differential pressure (21 bar), as the old loop was designed for a max flow rate of 120 m^3/h. The test results at ΔP 21 bar are reported in fig.5. The pump was running for more than 10 minutes at 100% GVF (fig.6). The test results show that the absorbed power remains constant, as well as ΔP; in the first 200 sec the gas flow rate gradually decreases from 78 m3/h to 60 m3/h, and then remains constant. This behaviour is probably due to some liquid still present in the pump and in the suction and delivery vessel. So at 1000 RPM, ΔP 21 bar and 100% GVF the pump has a flowrate of 60 m3/h (60% volumetric efficiency) and absorbed power is almost the same as at 0% GVF. The only measurement that indicates the 100% GVF is the increase of the liner temperature. Next objective is to repeat the test on NPV280 (medium size pump) on the new loop, at full speed and at ΔP up to 40 bar. Then the test will be repeated on NPV 1600 (DMBS pump), that will be able to reach a differential pressure of 70 bar. NPV1600 will be manufactured within June '99, and will be tested in the new multiphase loop as soon as will be completed by the manufacturers the FAT of electrical motor and frequency converter.

3.3. Pump mechanical seals

It has been decided to carry out a complete qualification test program for the mechanical seals, which are the most critical component in the pump, especially in subsea conditions.
The qualification program purpose is :
- to check the leakage of the new seal in reference conditions
- to check again the leakage of the seal after any test made to simulate some particular working conditions.

Qualification testing of mechanical seals started in July '98. It will be done on three different kinds of mechanical seals, and is scheduled to be completed by January '99.

It has been done simulating the vertical configuration and the lower seals condition, more critical due to the accumulation of sand and solid particles present in the process fluid.

The test bed for seals qualification is shown in fig.7. In short, it is a vertical structure, in which a shaft is rotated by an electrical motor, actuated by a frequency converter. The seal to be qualified is fitted on the end of this shaft. It is a double seal, and while the lower seal is in conctact with the atmosphere, the upper seal, that is the seal to be qualified , is in contact with a pressurized vessel in which any process fluid can be simulated. The seal is flushed and cooled by a clean mineral oil, at a pressure 2 bar higher than the process fluid pressure.
The main measurements are:
- oil inlet pressure and temperature
- oil outlet pressure and temperature
- process fluid pressure and temperature
- oil flow rate

The tests scheduled for every selected seal are:
- New seal leakage check (3 days continuous running, air as process)
- Test with typical well water in the process side (95,000 ppm salinity,10 days continuous running), to check the formation of scaling that can damage the seal
- Seal leakage check again (3 days continuous running, air as process)
- Test with 50% mineral oil, 50% salt water, 500 ppm sand in the process side (10 days continuous running), to check the typical mixture of a live well product, containing sand. Experience at Prezioso field has shown that solid particles are the main cause of damage to the mechanical seals, and mainly the lower seals, where solid particles settle, causing wear and encreased leakage.
- Seal leakage check again (3 days continuous running, air as process)
- N° 16 start-ups, with the final mixture; each start-up will be followed by a two-hour stop, to simulate a real shut-down and to give the sand time to settle around the seal
- Seal leakage check again (3 days continuous running, air as process)
- Final seal check (with disassembly and inspection)

- On the completion of qualification testing another short test will be performed with inverted differential pressure (i.e. process pressure higher than oil pressure), in order to to simulate the conditions that have been noted in Prezioso during some quick start-ups or sudden shut-downs
- Seal leakage check again, and again seal inspection.

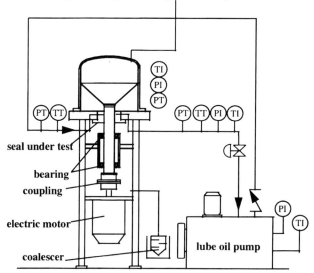

Fig.7: Simplified outline of the seal qualification test bed.

REFERENCES
[1] S.De Donno, G.Ferrari Aggradi, G.Chiesa : "Multiphase Production Technology Development Subsea Field Trials of a Pumping Unit Prototype". Proceedings of 7th DOT Conference, Montecarlo, Monaco, November 1993.
[2] G.Chiesa, S.De Donno, G.Ferrari Aggradi : "Subsea Testing of a Multiphase Pumping Unit" Proceedings of OTC Conference, Houston, May 1994.
[3] S.De Donno, P. Colombi, G.Chiesa, G.Ferrari Aggradi : "The Experience from Field Operation of a Subsea Multiphase Booster". Proceedings of OTC Conference, Houston, May 1995.
[4] P. Colombi, G.Chiesa, G.Ferrari Aggradi : "Multiphase Boosting: a Growing Technology for the Challenge of Economical Deepwater Developments". Proceedings of OMAE Conference, Firenze,Italy, June 1996.
[5] M.Granato, P.Colombi, G.Chiesa, C.Rossi, G.Ferrari Aggradi : "Field Operation of a Subsea Multiphase Boosting System". Proceedings of 6th International Conference on Multiphase Flow in Industrial Plants, Milano, Italy, September 1998.

C556/009/99

The application of computational fluid dynamics techniques for improved impeller design in re-rating centrifugal compressors

J A G SIERINK and **H T W HETTELAAR**
Thomassen International bv, Rheden, The Netherlands

SYNOPSIS

Rerating a centrifugal compressor is one of the options available for attaining process condition changes and/or obtaining an efficiency increase. An important aspect of any rerate is that the re-use of the existing casing(s) imposes dimensional constraints on the design. New impeller dimensions, diffuser lengths etc. are limited by the casing dimensions, providing an additional challenge for the designer to overcome.

This paper describes how modern Computational Fluid Dynamics (CFD) techniques can assist the designer in achieving this goal within the restrictions imposed. After an introduction, the general rerate design process is outlined. This is followed by a description of two gas flow related aspects that are assessed during the design process of centrifugal compressor impellers. The results of a CFD study are described where the influence of three geometric design parameters on the performance of a rerate impeller was investigated. Based upon these results, an improved impeller design was conceived, meeting the capacity and head requirements at increased efficiency within the dictated geometric limits.

This paper demonstrates the advantages and importance of employing modern design tools in the optimization process of an impeller design within dimensional constraints. It also illustrates how it may an important role in the ability to develop customized designs.

NOTATION

beta	Blade angle	
bbl	Blade to blade loading	
CFD	Computational Fluid Dynamics	
D_2	Impeller tip diameter	[m]
H	Total to total polytropic head	[J/kg]
Q	(Inlet) volume flow	[m³/s]
u_2	Impeller tip speed	[m/s]
1D, 3D	One dimensional, three dimensional	
η	Total to total polytropic efficiency	[%]
μ	Head coefficient = H / u_2^2	[-]
φ	Flow coefficient = $Q / (u_2 . D_2^2)$	[-]

1 INTRODUCTION

Centrifugal compressors are important parts of most chemical production plants. They are used to bring gas flows up to the pressure required for the different process parts. As such, these compressors are designed for a certain gas composition, volume flow (i.e. plant production capacity) and pressure ratio. Usually, a centrifugal compressor (or train of several compressors) is intended to run in the process for several decades. The required or desired process conditions of the plant, and therefore the design conditions of the compressor stages, usually change during this period. It can be decided to completely replace the existing compressors by new ones, in order to accommodate these changes. However, the existing compressor casings often have the capacity to allow for a change in operating conditions by means of installing new internal parts. This rerating offers the advantage that no changes are necessary to the compressor foundations and process piping. Compared to a compressor replacement, a rerate is normally a much less expensive solution to a change in conditions. It also offers the benefits of employing newly developed technology, like an increase in efficiency. In fact, the latter can be the primary goal of the rerate.

The design challenge to be overcome is that the rerate internals have to fit in the existing casing. This becomes even more important if the driving unit puts constraints on the applicable rotational speed. Because most rerates involve a capacity increase, it often means that the dimensions of the new internals are to be smaller than the dimensions the designer would have chosen in case of a new machine design. In order for the rerate to be a better choice than a compressor replacement, the performance of the new internals has to be optimized within the existing casing dimensions. In order to extend the experience to be able to perform these optimizations, a real size stage testing program was executed. The results of these tests have been incorporated in the design software that is used.

As another way to extend the experience, Computational Fluid Dynamics (CFD) offers the possibility to predict the compressor performance by simulation of the gas flow through its internal parts. This paper aims to show how general CFD design techniques can be of specific assistance to the rerate designer in the process of achieving an optimum design within the dimensional constraints imposed by the existing casing. To that end, it presents the results of a CFD study into the effects of three design parameters on the performance of a centrifugal compressor impeller, and how these results have been used to improve the design. The goal of

C556/009 © IMechE 1999

this study was to improve the efficiency of a rerate impeller designed for specific conditions within a specific available space. The efficiency of a compressor stage does not only depend on the design of the rotating impeller but also on the design of the stationary parts i.e. the diffuser, cross over bend and return channel to the next stage. The impeller was chosen as a starting point, for what is lost here can never be recovered in the stationary parts.

This paper consists of the following parts. It starts with some comments on the design process of a rerate impeller. The general process is outlined, followed by a short explanation on two gas flow related aspects one assesses during the design of an impeller. This background information is used in the presentation of the aforementioned CFD impeller improvement study, which forms the next and major part of this paper. The last part summarizes the conclusions that can be drawn from this study.

2 RERATE IMPELLER DESIGN

This section outlines the general rerate design process and some of its detailed aerodynamic aspects. It is meant to give background information on the next part that handles the CFD impeller improvement study.

2.1 Rerate design process
The general rerate design process starts by answering the question whether the casing(s) can accommodate the new process conditions. Especially in case of a capacity increase and/or decrease of the gas mole weight, the casing could be too small. This question is answered during the first phase in which the thermodynamic design of the internals is conducted. By means of a custom built thermodynamic computer model of the compressor, the main specifications of the internals are determined - the number of stages, the impeller flow coefficients, the required head per impeller, the diffuser geometry and so on. During this process, the experiences from previous jobs and tests are used to check the feasibility of the design. An important feasibility aspect is the space that is available for the individual compressor stages in combination with driver speed limits.

The second phase is the detailed design process of the internal parts. Mean streamline (1D) and quasi-3D flow analysis and design software[1] is used to achieve customized optimum stage designs within the space available, instead of using standard stages as is common practice for new machines. This gives the rerate design a performance (i.e. efficiency) advantage over a standardized new design. In case of an impeller design, the second design phase covers the determination of the inlet diameter, the number of blades, the exact blade angles and so on.

The feasibility issue of the first design phase may require second design phase activities. In case of comfortable stage conditions (i.e. head, flow, available space etc.), the attainability of a certain efficiency is not a real issue; it can be based on previous experience. Under more extreme conditions however, the attainable efficiency can only be based on the results of the detailed design process. In this case, the first and second design phase are performed simultaneously instead of sequentially.

---

[1] COMPAL (1D) and CCAD (quasi 3D), Concepts ETI, White River Junction, Vermont, USA.

The second design phase also includes the mechanical design aspects, such as impeller stress analysis. The rerate design process involves several other items like bearing and seal design, detailed driver assessment and rotordynamic analysis, which fall beyond the scope of this paper.

2.2 Aerodynamic design aspects
In this section, two aerodynamic aspects of rerate impeller design are discussed. The first subject is the blade to blade loading; the second concerns the secondary flow zones that appear at the impeller discharge.

2.2.1 Blade to blade loading
The task of the impeller as part of a centrifugal compressor stage is to increase the static and dynamic pressure of the gas flow. Partly, this is achieved by a reduction of the relative speed with which the gas flows through the impeller. The rate of this diffusion along the flow path is an important design parameter. It influences the growth rate of the boundary layers along the internal impeller surfaces. A locally too slow or too fast diffusion on the curved blades can result in boundary layers that are too thick or in early separation. These effects block or even stall the flow and have a negative influence on the impeller efficiency.

The rate of diffusion is expressed in terms of surface velocity distributions. It can be defined as the distribution of $\Delta W / \overline{W}$, where ΔW is the difference between the relative velocities of the gas on two opposite impeller channel surfaces and $\overline{W}$ their average value. When these two surfaces are a blade pressure and suction surface, the term blade to blade loading (bbl) is used. This refers to the relation between the blade relative velocity (i.e. pressure) difference ΔW, and the energy transfer from the blades to the gas. The largest boundary layers arise at the shroud sides of the blade suction surfaces (see section 2.2.2). Therefore, the shroud bbl is the most critical. Dallenbach (1) investigated the influence of the bbl on impeller performance. Japikse (2) describes a translation of these results into some empirical design rules regarding bbl distribution.

2.2.2 Impeller secondary flow zones
The gas flow out of each discharge area between two blades of an impeller can be divided into two zones. In this simplified model, the first zone is the primary flow of gas with a relatively high velocity and local efficiency. This is also referred to as jet flow and it covers the main part of the discharge areas. The second zone is the secondary or wake flow that is positioned near the shroud suction sides of the impeller blades and contains boundary layer gas with a low momentum and a low local efficiency. One of the investigations into secondary zone development at the impeller discharge can be found in Eckardt (3). The secondary zone is fed by a number of components/vortices that are indicated in figure 1 and can be listed as follows:

A. More flow is gathering at the approaching walls when it is forced to turn from an axial direction at the inlet to a radial direction at the discharge in its passage through the backward swept channels. Therefore, the flow is directed towards the hub pressure sides of the blades.
B. The change from an axial to a radial direction requires a pressure gradient from the outer towards the inner radius of the turn, corresponding to the velocity of the main flow. The flow in the boundary layers on the blade surfaces has a lower velocity than corresponds to this gradient and part of it is therefore pushed towards the inner radius of the turn.

C. In an absolute frame of reference, the flow particles are forced to follow a forward curved path because their path radius increases while they rotate with the impeller. The flow in the boundary layers on the hub and shroud surfaces have a lower velocity along this path and is therefore partly pushed towards its inner radius i.e. the blade suction surfaces. This is similar to the secondary flow effect producing vortex pair B.

D. In each of the inlet areas between two blades, the flow rotates against the impeller rotational direction due to its inertia. This counter vortex is transported through the channel while staying perpendicular to the main flow direction, which is radial at the impeller discharge.

E. A vortex similar to vortex C appears in the main (radial) part of the impeller. It lies in a plane that is perpendicular to the impeller rotation axis. At the discharge, this potential flow vortex tends to push the flow from the suction surface towards the pressure surface.

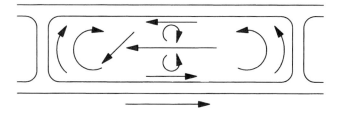

Figure 1: Shrouded impeller discharge (in radial inward view) showing secondary flow zone components. Free reproduction from Elder (4).

The secondary flow zone is an important aspect of the impeller design process. Much research has been done in order to determine its origin and its influence on impeller performance.

For an overview of this research and a translation of the results into a basis for a practical design tool, the reader is referred to Japikse (2). For this paper, the most important aspect to note is that the secondary zones at the shroud suction sides of the impeller blades can be thought of as being the areas that comprise the aerodynamic losses of the impeller.

3 USING CFD TECHNIQUES TO IMPROVE AN IMPELLER DESIGN

An impeller design may be improved by changing its geometric parameters. With the aim to understand more about the influence of these parameters on the performance, a study has been performed. This section presents the results of this study.

3.1 Geometric parameter study

The goal of the detailed design process (the second design phase of section 2.1) is to obtain an optimum design for the given conditions. How to achieve an optimum efficiency by changing the stage geometry is a question that can be partly answered by testing. Another method is to use CFD to assess the influence of geometric parameters on the performance. At present, the results of CFD calculations are still not completely reliable. Absolute values that have been obtained by using CFD have to be evaluated with care. CFD may however effectively be used to compare calculated results to each other.

The next two sections describe some results of a geometric parameter study, based on the use of CFD techniques. The study was aimed at improving a specific impeller design in order to extend the general knowledge about the influence of geometric parameters on impeller performance. The improvement of an impeller was chosen as a subject, because it is the main and first part of a compressor stage.

The subject impeller has been designed for a head coefficient of 0.6, a flow coefficient of 0.064 and a limited available space in both the axial and radial direction. Starting with this shrouded reference design, several alternative designs have been developed. These designs differ from each other and from the reference design in (the value of) one of the geometric parameters. The overall dimensions (diameter and axial length) of the designs have been kept constant. They act as the dimensional constraints normally encountered during a rerate impeller design. The gas flow through all alternative designs has been modelled by use of a CFD package[2]. The results of these CFD calculations have been used to calculate the mass averaged head and efficiency of the impeller designs. Together with views of the calculated flow fields, this allows for assessment of the influence of the geometric parameter changes on the impeller performance.

3.2 Parameter study results
The influence of five geometric parameters has been investigated. Three of them are discussed here. These are inlet wrap angle, blade angle distribution and discharge rake angle.

For manufacturing reasons, the impeller is designed such a way that the blade surfaces are ruled surfaces. As such, the designer can only control the blade angle (beta) distribution along the shroud. The resulting beta distribution along the hub is calculated by the design software program in order to comply with the ruled surface design. The reference impeller has a standard S-shaped shroud beta distribution with a rather extreme curvature of the hub beta distribution towards the trailing edge (see Figure 2). The latter can have a negative influence on the boundary layer build up and therefore on the impeller performance. A less curved hub beta distribution would require a longer blade contour along the hub. This can be achieved by increasing the axial length and/or the diameter of the impeller, which is not allowed here since the available space is fixed for the study.

Another way of achieving a longer hub contour is the application of an inlet wrap angle. This angle can be identified in the axial view of the impeller inlet. Figure 8 shows an impeller with a positive inlet wrap angle W. The points where the blade leading edges meet the hub have been rotated W° in the rotational direction. This angle is measured relative to the radial lines through the rotation axis and the points where the blade leading edges meet the shroud.

Inlet wrap angles of 5°, 10° and 15° have been applied as a modification to the reference design that has an inlet wrap angle of 0°. Figure 2 shows that these designs have a decreasing curvature of the hub beta distribution, while the shroud beta distribution is not changed.

---

[2] W.N. Dawes' BTOB3D code, Whittle Laboratory, Cambridge, UK

C556/009 © IMechE 1999

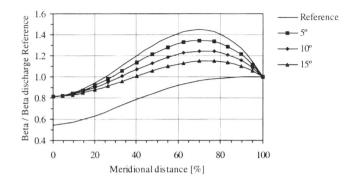

Figure 2: Influence of investigated inlet wrap angles on blade angle distribution (constant shroud blade angle distribution).

Table 1 summarizes the performance results of the investigated impeller designs. It can be seen that, for the reference impeller design, the efficiency increases with inlet wrap angle, while the head decreases. A physical explanation for these effects is not discussed here. However, it can be noted that the less curved hub beta distributions of the inlet wrap angle designs result in somewhat lower bbl's along both the hub and shroud contour.

Table 1: CFD calculated performance results of investigated impeller designs.

Impeller design	η – η Reference [% point]	H / H Reference [-]
Reference	0.0	1.00
Inlet wrap angle 5°	0.2	0.97
Inlet wrap angle 10°	0.3	0.95
Inlet wrap angle 15°	0.5	0.93
Blade angle distribution 1	0.0	0.99
Blade angle distribution 2	-0.1	1.00
Blade angle distribution 3	-0.1	1.00
Discharge rake angle 15°	0.0	0.99
Discharge rake angle 30°	0.2	0.98
Discharge rake angle 45°	0.8	0.95
Improved	3.0	1.00

A second geometric parameter that has been investigated is the blade angle distribution. As indicated, only the shroud beta distribution is changed and the hub beta distribution is derived from this. The standard optimum shroud beta distribution is S-shaped; see, for instance, Joslyn (5). The modern approach of basing the shroud beta distribution on the bbl along this contour has been used in the parameter study. Figure 3 shows three investigated blade angle distributions. Going from distribution 1 to 3, the major shroud beta change is displaced towards the blade trailing edges, thereby decreasing the curvature of the hub beta distribution.

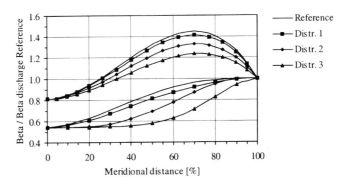

Figure 3: Investigated blade angle distributions.

Figure 4 shows the resulting shroud bbl's, as calculated by the quasi-3D design software program. It can be seen that the relative bbl follows the displacement of the major shroud beta change towards the trailing edges, resulting in a lower relative bbl upstream. This means that the major energy transfer from the impeller blades to the gas is shifting downstream as well. It would be expected that these rather large changes in bbl have a significant effect on the impeller performance. Table 1 shows that this is not the case here. The efficiency and head are not influenced by the investigated changes in the beta distributions.

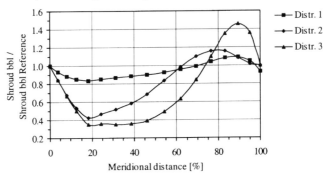

Figure 4: Shroud blade to blade loading for investigated blade angle distributions.

This insensitivity indicates that other factors are more important. Therefore, the described influence of the inlet wrap angle on the bbl cannot be the only explanation for its effect on the impeller performance. A more detailed study of the CFD results suggests that the influence of the geometry changes on the impeller inlet and outlet flow patterns offers a better starting point for explaining the changes in performance. This assessment method is used in the following discussion.

The application of an inlet wrap angle is not the only method to elongate the blade hub contours in order to decrease their beta distribution curvature. It can also be achieved by applying a discharge rake angle. Figure 8 shows an impeller with a positive rake angle R. The points where the blade trailing edges meet the hub have been rotated R° against the rotational

C556/009 © IMechE 1999

direction. This angle is measured relative to the axial lines through the points where the blade trailing edges meet the shroud. These axial lines run parallel to the impeller rotation axis.

Usually, rake is applied to decrease the trailing edge stresses of unshrouded impellers, generated by the centrifugal forces acting on the backward swept blades; see, for instance, Cumpsty (6). For the parameter study, attention was focussed on its influence on the performance. Rake angles of 15°, 30° and 45° have been applied as a modification to the reference design, having a rake angle of 0°. The resulting blade angle distributions are similar to those of figure 2; the only difference is that the decrease in the curvature of the hub beta distribution is less. Amongst others, the use of discharge rake also results in a slight decrease of the bbl along the hub and shroud.

Table 1 shows that the performance influence of discharge rake on the reference impeller design is similar to that of inlet wrap. The efficiency increases with the rake angle while the head decreases. Figure 5 shows the origin of the efficiency increase. The use of discharge rake decreases the secondary zone area of low local efficiency (near the shroud suction sides of the blades) relative to the primary zone area of higher local efficiency. The intensity of the secondary zone increases with the rake angle. This intensification has no negative effect on the calculated mass averaged efficiency because the local absolute meridional discharge velocity Cm is small. On the other hand, the increase with discharge rake of the primary zone area has a large positive effect on the efficiency because of the higher local Cm.

It can be concluded that the efficiency of the reference impeller design can be increased within the same overall dimensions by applying an inlet wrap and/or a discharge rake angle to the blades. However, this has a negative effect on the delivered head. The reference impeller design does not seem to be very sensitive to changes in the bbl distribution.

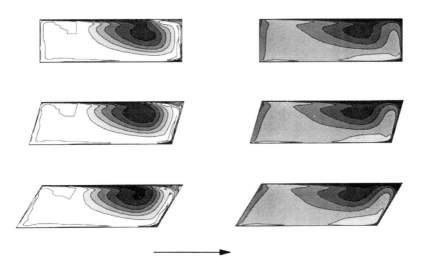

Figure 5: Radial inward views of discharge area between two blades, showing efficiency and absolute meridional discharge velocity distributions of reference and two discharge rake angle impeller designs. A darker colour indicates a lower value.

3.3 Improved impeller design

The results of the parameter study have been used to improve the reference impeller design. Amongst others, this improved design has an inlet wrap angle of 15° and a discharge rake angle of 45°. The resulting head loss has been compensated by a decrease of the discharge blade angle; i.e. the blades are more radial towards the trailing edges. Figure 6 shows the resulting beta distribution.

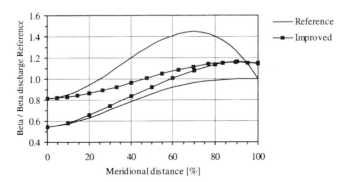

Figure 6: Blade angle distribution improved impeller design.

Table 1 indicates that the improved impeller design has a 3 % point higher calculated efficiency than the reference impeller, while delivering the same head at the same flow within the same overall dimensions. Figure 6 shows that the beta distributions along the hub and shroud contour are more radially directed, quite similar and not very curved. This results in a favourable distribution of the primary and secondary flow zones at the impeller discharge, as can be seen in figure 7. Compared to the reference design (see figure 5), the improved design has a smaller secondary zone area that is pushed more towards the shroud suction sides of the blades. The positive influence of the Cm distribution on the mass averaged efficiency (as indicated in the discussion on figure 5) is also evident here.

Figure 7: Radial inward views of discharge area between two blades, showing efficiency and absolute meridional discharge velocity distributions of improved impeller design. A darker colour indicates a lower value.

Figure 8 shows a wire frame image of the improved design. It should be noted that the described kind of CFD investigation is only a first step towards a completed design for an improved impeller. A characteristic of the improved design is a reduction of the discharge blade angle, resulting in a reduction of the stable operating range. This has to fit into the intended service of the compressor. Furthermore, the mechanical feasibility of the design

C556/009 © IMechE 1999

must be checked (stresses, natural frequencies etc.). The last step is a full scale test of the design in order to determine the actual performance improvement. The discussion on the parameter study shows however, that the use of CFD techniques is a useful tool for taking the first step. It offers a relatively fast method to test ideas and gain a more sound insight into the factors that determine the performance of an impeller.

Figure 8: Improved impeller design (shroud not shown).

CONCLUSION

General CFD design techniques offer a useful tool for the specific improvement of a rerate impeller design within the limited space available. It has been shown that these tools can help the designer in gaining more knowledge about the factors that determine the impeller performance. An optimum impeller design is a prerequisite for an optimum design of a compressor stage. The geometric parameter study that has been presented has produced some very practical results. A relatively large improvement of the calculated efficiency can be achieved by means of some relatively small design changes. Furthermore, the study shows that changes in design parameters do not necessarily lead to the normally expected changes in impeller performance.

A critical attitude towards these kinds of studies is necessary. The calculated performance improvement has to be confirmed during the rest of the compressor design process. This includes the mechanical and operational feasibility of the design. However, when used wisely, CFD techniques offer a relatively fast and cheap method for taking the first step in the optimization process of a compressor design.

In general, it can be concluded that the rerate problem of a limited space can be partly overcome by using modern design tools to perform an accurate aerodynamic design of the new internal parts. This way, a customized design can be conceived that has a near to optimum performance (i.e. efficiency) under the given conditions.

REFERENCES

(1) Dallenbach, F., The aerodynamic design and performance of centrifugal and mixed-flow compressors, *Centrifugal compressors*, SAE Technical Progress Series, Vol. 3, 610160, 1961.

(2) Japikse, D., *Centrifugal compressor design and performance*, Concepts ETI, White River Junction, Vermont, USA, 1996.

(3) Eckardt, D., Flow field analysis of radial and backswept centrifugal compressor impellers, Part 1: Flow measurements using a laser velocimeter. In Gopalakrishnan, S., Cooper, P. and others (editors), *Performance prediction of centrifugal pumps and compressors*, ASME, 1980.

(4) Elder, R.L., Forster, C.P., *Lecture notes centrifugal compressors*, Cranfield Institute of Technology, Cranfield, UK, 1986.

(5) Joslyn, H.D., Brasz, J.J., Dring , R.P., *Centrifugal compressor impeller aerodynamics: an experimental investigation*, ASME paper no. 90-GT-128, 1990.

(6) Cumpsty, N.A., *Compressor aerodynamics*, Longman, Essex, UK, 1997.

C556/014/99

Process centrifugal compressors – latest improvements of efficiency and operating range

K LÜDTKE
GHH BORSIG Turbomaschinen GmbH, Berlin, Germany

SYNOPSIS During the past decade a number of evolutionary improvements were introduced to enhance compressor aero-thermodynamics. Since energy consumption and operational flexibility hold a high rank on the priority list numerous efforts were made to increase both efficiencies and operating ranges for multistage singleshaft centrifugal compressors; as predominantly used in the hydrocarbon processing industry.

NOTATION
Symbols not explained in text :

a	axial clearance impeller/diaphragm	Ra	arithmetic centreline roughness
a_0	sonic velocity at impeller inlet	s	work input coefficient $\Delta H/u_2^2$
AR	conical diffuser area ratio	u_2	impeller tip speed
b_2	impeller exit width	V	actual volume flow at inlet
c	gas velocity	ΔH	enthalpy difference (Euler head)
d_2/r_2	impeller outer diameter/radius	ψ	pol. head coefficient $h_p/(u_2^2/2)$
L/d_1	conical diffuser length/inlet diameter	ζ	head loss coefficient
m	mass flow	η_3	dynamic viscosity impeller exit
M_{u2}	tip speed Mach number u_2/a_0	φ	flow coefficient $4V/(\pi d_2^2 u_2)$
Re	Reynolds number $u_2 b_2/\nu_1$	ν_1	kinematic viscosity impeller inlet
r_5	centre of gravity radius volute exit area	ρ_3	density impeller exit

1. INCREASE OF EFFICIENCY

1.1 Optimization of aero-thermodynamics
Centrifugal compressor stages achieve the highest possible efficiency when designed with a flow coefficient φ of around 0.07 to 0.10; assuming the application of impellers with twisted (so-called 3D-)blades.

As a matter of fact the volume flow reduces towards the rear end of the machine and the diameters of subsequent impellers on one single shaft cannot be stepped down in diameter in order to obtain optimum flow coefficients for all stages. So, necessarily, the efficiency reduces for each subsequent stage for single-shaft compressors.

Figure 1 contains the η-φ-ranges of two multi-stage compressors from the first to the last stage; the first is a conventional and the second is a high flow coefficient design resulting in 20 percent smaller impeller diameters, 25 percent higher rotational speed and five percent lower power consumption.

During the last decade high flow coefficient impellers with high hub/tip ratios were developed. Three constraints have to be observed, though. Rotordynamic criteria for these multistage units are of utmost importance : in order to avoid subsynchronous vibrations the shaft stiffness ratio, i.e. rigid support critical over maximum speed N_{c1rig}/N_{max}, must be greater than 0.4 to 0.6; depending on the average gas density, as stated by Fulton (1). Therefore, since high flow coefficient impellers have smaller shaft diameters and greater axial lengths, the maximum number of impellers to be accommodated in one casing is smaller than for conventional designs; trends are shown in Figure 2. So it might well be that one ends up with a high efficiency requiring two casings vs. a single casing with medium efficiency. Normally, the increased costs are not justified by a power reduction in this order of magnitude. The second constraint is the increased rotational speed that may exceed acceptable limits dictated by maximum driver or gear speeds or by experience or by customary practice especially for small high pressure compressors. So far the maximum multi-stage singleshaft compressor speed in the author´s company is around 20,000 rpm. The third constraint of high flow coefficient impellers is the increased stress level at given tip speed of approximately 115 percent compared to conventional impellers. This leads to a higher material yield strength to maintain stage head or, in case of a limited material yield strength (e.g. if H_2S is present), to a reduced stage head.

1.2 Increased diffuser diameter ratio

The „diaphragm package" downstream of the impeller -normally consisting of a vaneless diffuser, a crossover bend and a return vane channel- serves as a means a) to convert the high kinetic energy at the impeller exit into static pressure and b) to deswirl the flow before it enters the next impeller. As odd as it may sound: the longer the flow path in the diffuser/return vane unit the higher the total-to-total stage efficiency (suction s to discharge d, Figure 3) as long as diffuser diameter ratios of around 1.8 are not exceeded. Lindner (2), a former colleague, reported on test results which were obtained with the author´s company´s compressor stages gaining approximately 2.5 percentage points efficiency when changing the diffuser ratio from 1.45 to 1.65 of a 3D-bladed impeller with a flow coefficient of 0.09 at tip speed Mach numbers of 0.55 to 0.93; configuration as in Figure 3. Since the total-to-total efficiency of impeller+diffuser (stations s to 4, Figure 3) must go down for an extended diffuser the reason for the improvement can only be the return vane channel. At considerably reduced inlet kinetic energy, the channel responds with lower head losses $\Delta h_{5d}=\zeta c_5^2/2$ (nomenclature see Figure 3); the loss coefficient ζ is reduced additionally, as the deswirling is distributed over a longer flow path.

In (2) it was also reported that the low diffuser ratio yields a lower head rise to surge with the horizontal tangent on the pressure/volume curve being closer to the rated point. Thus contributing to a lower flexibility of the compressor to cope with changes of the molecular mass; often the case in the oil and gas industry. High incidence losses at the return vane leading edges during partload are presumably responsible for this inadequate curve slope.

Therefore, long diffusers are definitively the better solution when discussing efficiencies and operational flexibility.

Table 1 shows geometric and aerodynamic parameters, as well as shop test results of two similar three-stage contract compressor sections with medium flow coefficients and high Mach numbers. The increase of the average diffuser ratio from 1.52 to 1.67 resulted in an efficiency increase of 3.3 percentage points. In order to be able to compare efficiencies they were adjusted for the different heat radiation losses, diameters and impeller exit width/diameter ratios according to proven empirical experience and also for Reynolds numbers according to the method, described in (7), which was introduced by the „International Compressed Air and Allied Machinery Committee" (ICAAMC). The cross-sections of the relevant compressors are shown in Figure 4.

Table 1. Two performance tested compressor sections with different diffuser ratios and medium flow coefficient impellers

Version	A	B
code	Ampolaus 1st section	Shedaus 1st section
number of stages	3	3
standard impeller type	90.7 60.7 40.7	80.7 50.7 35.8
impeller diameter d_2	460; 460; 430	562; 552; 517
width/dia ratio b_2/d_2	0.0585; 0.0406; 0.034	0.0496; 0.0373; 0.0343
diffuser ratio r_4/r_2	1.49; 1.49	1.66; 1.69
volute type / area	external/ circular, oval	external / circular, oval
standard volute size; r_5/r_2	600-2; 1.57	600-2; 1.67
average diffuser ratio	1.52	1.67
flow coefficient φ, optimum	0.092-0.045	0.076-0.042
tip speed Mach number M_{u2}	0.98-0.88	1.00-0.80
avg. test Re number Re_t	2.2×10^6	0.13×10^6
pol. stage eff. η_p, measured	0.810	0.836
adjusted for radiation,	-	
d_2, b_2/d_2, Re $\eta_{p\,adi}$		0.843
efficiency difference $\Delta\eta_p$		0.033

Table 2 shows that low flow coefficient stages can also be improved in the efficiency by applying longer vaneless diffusers. The very similar three- and four- stage contract compressor sections, having the same standard impeller types, differ in their average diffuser ratios (1.65 versus 1.48); bringing about an efficiency difference of 2.4 percentage points in favour of the long diffuser section. Again, measured efficiencies were empirically adjusted to make them comparable. It should be added that version D had a smaller axial clearance between impellers and diaphragms of $a/r_2 = 0.015$ versus 0.035 (version C) which contributed somewhat to the efficiency improvement (for details see 1.5). The cross-sections are shown in Figure 5. As can be derived from the test results, the findings of the development tests (2) are basically confirmed: some two to three percentage points difference in the efficiency for stages with increased diffuser radius ratios.

Table 2. Two performance tested compressor sections with different diffuser ratios and low flow coefficient impellers

Version	C	D
code	Pusind, 2nd section	Shebru, 3rd section
number of stages	3	4
standard impeller type	10.82	10.82
impeller diameter d_2	615;560;515	515;505;480;455
width/dia ratio b_2/d_2	0.015;0.0148;0.015	0.0149;0.0133;0.0129;0.0136
diffuser ratio r_4/r_2	1.51; 1.46	1.59; 1.62; 1.70
volute type / area	external / circular	external / circular, oval
standard volute size; r_5/r_2	600-5; 1.46	600-45; 1.69
average diffuser ratio	1.48	1.65
flow coefficient φ, optimum	0.0125	0.012
impeller/diaphragm gap a/r_2	0.035	0.015
tip speed Mach number M_{u2}	0.85-0.61	0.68-0.55
avg. test Re number Re_t	1.5×10^5	3.8×10^5
pol. stage eff. η_p, measured	0.639	0.673
adjusted for d_2, b_2/d_2, Re	-	0.663
efficiency difference $\Delta\eta_p$		0.024

1.3 Improved volute geometry

The „diaphragm package" of a final stage at the end of a section consists of an (normally vaneless) annular diffuser, a volute and a conical exit diffuser. The volute collects the gas circumferentially at the annular diffuser exit and guides it into the conical diffuser and discharge nozzle. All three elements serve the purpose of efficiently converting a great deal of the kinetic energy at the impeller exit into static pressure. One casing can have up to four of these packages, i.e. four sections with three intercoolers.

In Table 3 three shop performance tested single stage contract compressors with different volute configurations are compared with each other. They have the same scalable standard impellers, two of which were slightly modified regarding the exit width. All three tip speed Mach numbers are in a range where Mach number differences do not affect the efficiency. To make them comparable, the efficiency of version E was adjusted to radial inlet and version F was adjusted to zero cutback.

Table 3. Three performance tested singlestage compressors with different volutes

Version	E	F	G
code	Rash 1	Lusab 1	Saiboost
rotor type	overhung	beam	beam
inlet	axial	radial	radial
standard impeller type	90.8	90.8	90.8
impeller disk diameter d_{20}	815	475	486
impeller blade diameter d_2	815	460	486
width/dia ratio b_2/d_2	0.0664 [1)]	0.080 [2)]	0.0631 [3)]
annular diffuser type	p a r a l l e l w a l l e d , v a n e l e s s		
annular diffuser ratio d_4/d_2	1.4	1.61	1. 94-1.5, avg. 1.72
volute type	external	internal	internal
volute area	circular/oval	rectangular	rectangular

volute outer radius/imp. radius r_a/r_2	2.2	2.06	2.06
volute axial length/impeller dia l/d_2	0.31	0.42	0.66
effective diffuser ratio r_5/r_2	1.72	1.61	1.77
exit diffuser L/d_1, AR	4.04 / 1.84	1.55 / 1.37	3.98 / 2.10
flow coefficient φ, optimum	0.084	0.094	0.096
tip speed Mach number M_{u2}	0.35	0.56	0.52
pol. stage efficiency η_p, measured	0.87	0.825	0.855
adjusted to radial inlet & $d_2=460$	0.83		
adjusted to zero cutback		0.815	

[1] Original standard impeller   [2] Std. impeller exit width increased   [3] Std. impeller exit width decreased. The impeller is shrouded and has backward curved 3D-blades.

Performance curves are directly compared in Figure 6.

Version E is the traditional textbook volute, one-sided offset, developing externally from the diffuser exit. Ten different sizes of these volutes per compressor frame size use to be part of the modular design system. Stage efficiencies are satisfactory and axial lengths are small to cope with rotordynamic requirements of multi-stage machines. However, the annular diffuser ratio is relatively low and the volute tongue is close to the impeller exit causing losses due to an uneven circumferential static pressure distribution. The radial space reqirements are excessive and the circular cross-sections allow casting as the sole manufacturing method. Version F has an acceptable annular diffuser ratio and a reduced outer radius, however, the reduced effective diffuser ratio results in additional losses through reacceleration of the flow. Version G combines the advantages of E and F: placement of the tongue further away from the impeller (increase of d_4/d_2), and increase of the effective diffuser ratio. The efficiency is increased because :

a) the circumferential static pressure distribution at the impeller exit is excellent in spite of the variable ratio of the annular diffuser (as revealed by a viscous three dimensional CFD calculation)

b) the average throughflow velocity level in the volute is reduced

c) the swirl velocity (responsible for the bulk of the losses) does not play a significant role due to the diminished meridional velocity in the long diffuser.

The high efficiency, the outstanding axial and radial adaptability to various geometric requirements and the options for milling and welding (by avoiding any circular cross-sections) are good reasons to replace version E volutes by version G volutes.

1.4 Reduction of flow channel surface roughness

Casey (3) suggested to define an equivalent pipe flow having a friction factor similar to the compressor flow. Using the well-known Moody diagramme, giving friction coefficient λ vs. Reynolds number $Re=u_2b_2/v_1$ with the relative arithmetic average roughness Ra/b_2 as a parameter, an approximation of the influence of the surface roughness on the efficiency can be calculated as follows :

Efficiency change with friction coefficient :

$$\Delta\eta_p = -\frac{c}{s}\Delta\lambda \quad .$$

The Colebrook-White formula for the friction factor was first published by Moody (4) :

$$\lambda = \cfrac{1}{\left[1.74 - 2\log_{10}\left(2\cfrac{Ra}{b_2} + \cfrac{18.7}{Re\sqrt{\lambda}}\right)\right]^2} \qquad .$$

Reynolds number :

$$Re = \cfrac{M_{u2} a_0 \cfrac{b_2}{d_2} d_2}{v_1} \qquad .$$

The equation was calibrated by utilizing test results that were systematically conducted on various stages with known roughness and Reynolds number. Therefore the dimensionless proportionality factor c as a function of b_2/d_2 and M_{u2} is defined as follows:

$$c = \frac{\left(1 - \eta_p\right)s}{\lambda} \qquad .$$

c is between 6.0 and 8.0 for wide impellers and between 8.0 and 12.0 for narrow impellers (low values at low Mach numbers). As can be seen from Figure 7, which shows as an example how the efficiency increases when the flow path roughness is reduced to 50% : the efficiency picks up between one and three points depending on the relative impeller exit width and the tip speed Mach number. So there is a great potential to improve low flow coefficient stages; typical for many applications in the oil and gas industry.

Surface smoothing is carried out conventionally and also by chemical or electropolishing. Needless to mention that it does not make sense to treat compressors handling gases with contaminating constituents.

1.5 Minimizing impeller / diaphragm clearance

The magnitude of clearances between the impeller hub and cover disks and the opposite diaphragm walls has an influence on the losses caused by disk friction, as revealed by Daily and Nece(5). The frictional power of a two-sided disk is :

$$P_F = C_M \frac{\rho_3}{2} u_2^3 r_2^2 \qquad .$$

Torque coefficient and Reynolds number are :

$$C_M = \frac{0.102\left(\cfrac{a}{r_2}\right)^{0.1}}{Re^{0.2}} \, ; Re = \frac{\rho_3 u_2 r_2}{\eta_3} \qquad .$$

C_M is valid for turbulent flow and separate boundary layers on both sides of the clearance. The disk friction loss per unit of gas power of a compressor stage is :

$$\frac{P_F}{P_i} = \frac{C_M \rho_3 u_2^3 \left(d_2^2 / 4\right)}{2 m s u_2^2} \qquad .$$

Substituting mass flow m with $V_3\rho_3$ and introducing the flow coefficient φ leads to a very simple equation for the impeller disk friction loss :

$$\frac{P_F}{P_i} = \frac{C_M}{4\pi s\,\varphi\left(V_3\,/\,V_{0t}\right)} \qquad .$$

The „feeling" that a disk, rotating in a denser gas and at a higher speed, will consume more frictional power P_F is not betrayed by this equation because the gas power P_i increases accordingly.

The torque coefficient, depending on the Reynolds number, is between 0.002 and 0.008 for most applications. The total to static impeller volume ratio V_3/V_{0t}, derived by the author (6), is a strong function of the Mach number and is between 0.7 for heavy gases and ≈1.0 for H_2 rich gases.

Figure 8 clearly shows the influences of the various parameters and the possible improvement potential of a reduced clearance ($a/r_2=0.02$ vs. 0.10). So the efficiencies of very low flow coefficient stages (φ=0.01) can be improved by an impressive one percentage point by reducing the axial gap from 10 to 2 percent of the impeller radius. One half percentage point improvement is possible for impellers with a flow coefficient of φ=0.025. For very high flow coefficient impellers of φ>0.1 the improvement is less than one tenth of a percentage point.

1.6 Efficiency increase through rub-resistant labyrinth seals

Leakage flows occur in all compressor labyrinth seals : at impeller cover disks, interstage shaft sections, buffergas seals, and at the balance piston. These parasitic flows, which have to be continuously compressed, do not contribute to the process mass flow. They account for three percent to, as high as, twelve percent of the overall power depending on flow coefficient level, pressure ratio and impeller arrangement.

Especially compressors with narrow impellers arranged in-line can be made more efficient if the conventional steel labyrinths are replaced by „plastic" seals with clearances of only 30 to 40 percent of steel seals. This "plastic engineering" is introduced more and more into today`s process compressors.

There are basically two types of plastic seal materials :

• Abradable materials, like polytetrafluoroethylene (PTFE) combined with mica and silicon-aluminum(AlSi) combined with polyester (PE), permit rubbing of rotating parts by giving way to intruding tips through abrasion, i.e. adapting their geometry locally. Labyrinth tips are of conventional material (e.g. steel).

• Rub-resistant materials forming the labyrinth tips, like polyamide-imide (PAI) and poly-ether-ether-ketone (PEEK), tolerate rubbing by deforming locally and resuming nearly the former shape. AlSi+PE can be used on stationary as well as on rotating parts; all others are used as stationary parts.

The overall energy savings attainable by applying rub-resistant labyrinth seals is between 2 and 8 % and varies with compressor size, impeller flow coefficient, impeller arrangement and pressure ratio.

2. INCREASE OF OPERATING RANGE

Multi-stage process compressors normally equipped with parallel-walled vaneless diffusers downstream of the impellers improve the surge limit by diffuser modification. The underlying thought is to stabilize diffuser flow right after the impeller.

2.1 Diffuser pinching

As Senoo and Kinoshita (8) pointed out, there is a critical (vaneless) diffuser inlet flow angle (from tangent), below which the flow reverses causing rotating stall and eventually triggering the incipience of surge. This is due to the drastically reduced radial velocity component at the front wall of the diffuser not having sufficient kinetic energy to maintain its forward movement to overcome the radial pressure gradient. Diffuser pinching increases the flow angle at the rated point, shifting the critical flow angle to lower flows and thus widening the operation range by warranting stable flow conditions over a wider capacity domain, as can be seen from Fig.9. The experiments conducted and reported by the author (9), yielded a surge flow reduction of 27 percent when replacing the parallel-walled by a highly tapered diffuser. For this extreme case of diffuser pinching the head reduction and hence the efficiency trade off is some five percent. So in principle turndown improvement is not free of charge. However, a very good compromise can be achieved by a constant area diffuser which improves the surge point by approximately ten percent at a cost of only one percent efficiency decrease, as described in (9). Also a mild diffuser taper of 10 to 15% downstream of the medium to high flow coefficient impeller and then continuing on with a parallel walled diffuser is quite effective at no appreciable efficiency decline.

2.2 Low solidity diffuser vanes

Senoo et al. (10) reported on the favourable effect of low solidity diffuser (LSD) vanes located downstream of the impeller. These cambered vanes with a great pitch/length ratio, having no throat whatsoever between adjacent vanes, stabilize diffuser flow substantially by imposing non-critical angles on the diffuser flow at partload. Additionally avoiding the adverse effects of conventional normal solidity diffuser vanes, whose throats have a strong tendency to choke at little more than the design flow. Figure 10 shows results of measurements on a medium Mach number multi-stage process compressor with and without LSD vanes on each stage. The surge flow was improved by 28 %, the head rise to surge was improved from 110% to 117% (beneficial for molecular mass changes and controllability), the rated point efficiency remained constant, the part-load efficiency was improved and the choke flow was reduced from 142% to 133%. The overall curve range V_{choke}/V_{surge} was substantially increased from 2.2 to 2.9.

The effect of the LSD vanes is basically the same as diffuser pinching : increase of flow angles (from tangent) at part load so that no surge triggering reverse flow can take place where the vaneless diffuser normally stalls. After a thorough calculation of flow angles along the performance curve the vanes must be positioned in such a way that, at the desired part load and at the rated point, the incidence angle $i = (\alpha_{chord} - \alpha_{flow})$ is positive; assuring a positive lift coefficient and a sufficiently small drag coefficient. Thus diffuser flow angles are increased by the vanes. This is also the case from rated to an appropriate volumetric overload. At higher overloads the incidence angle and the lift coefficient become negative (i.e. flow angles are reduced by the vanes) and the drag coefficient increases dramatically so that the LSD vanes shift the vaneless choke point to smaller volume flows. The chord angle is between the logarithmic spiral chord (resulting from conformal mapping from linear to angular) and the circumferential direction of the cambered profile.

   C556/014 © IMechE 1999

3. CONCLUSIONS

Some of the latest performance improvements in the field of aero-thermodynamics of process centrifugal compressors are described.

The overall compressor efficiency could be increased by developing high flow coefficient impellers with high hub/tip ratios, to be used as first stages, that avoid narrow low efficiency impellers at the discharge end of these multistage singleshaft compressors.

An increased radius ratio of the vaneless diffuser downstream of the impeller resulted in a higher stage efficiency not only with high but also with low flow coefficient stages.

Final stages of compressor sections, having volutes, could be augmented in their efficiency by developing a novel annular diffuser / rectangular scroll / conical diffuser package that is by far superior to other designs.

Low flow coefficient stages (i.e. with very narrow flow channels) have a relatively high enhancement potential by reducing flow path surface roughness, minimizing the clearance between impeller and diaphragm and applying rub-resistant labyrinth seals.

Diffuser flow angle steepening, through width pinching or installing LSD vanes, is a powerful means to stabilize the flow and thus extending the operating range considerably.

4. REFERENCES

(1)Fulton, J.W. „The Decision to Full Load Test a High Pressure Centrifugal Compressor in its Module prior to Tow-out", IMechE Conference Publications, 1984.

(2)Lindner, P. „Aerodynamic Tests on Centrifugal Process Compressors-Influence of Diffusor Diameter Ratio, Axial Stage Pitch, and Impeller Cutback", Transactions of the ASME, Journal of Engineering for Power, Vol. 105, No. 4, pp. 910-919, 1983.

(3)Casey, M.V., "The Effects of Reynolds Number on the Efficiency of Centrifugal Compressor Stages", ASME Paper No. 84-GT-247, 1984.

(4)Moody, L.F., "Friction Factors for Pipe Flow", ASME Transactions, 1944, 671-684.

(5)Daily, J.W., and Nece, R.E. „Chamber Dimension Effects on Induced Flow and Frictional Resitance of Enclosed Rotating Disks", Transactions of the ASME, Journal of Basic Engineering, March 1960, pp.217-232.

(6)Lüdtke, K., „Twenty Years of Experience with a Modular Design System for Centrifugal Process Compressors", Proceedings of 21st Turbomachinery Symposium, 1992, Texas A&M University, pp.21-34.

(7)Strub, R.A.et al, „Influence of the Reynolds Number on the Performance of Centrifugal Compressors", ASME Paper 87-GT-10, 1987 and Transactions ASME, Journal Turbomachinery 109, 1987, pp. 541-544.

(8)Senoo, Y., and Kinoshita, Y., 1977, "Influence of Inlet Flow Conditions and Geometries of Centrifugal Vaneless Diffusers on Critical Flow Angle for Reverse Flow", Transactions of the ASME, Journal of Fluids Engineering, 1977, pp. 98-103.

(9)Lüdtke, K., "Aerodynamic Tests on Centrifugal Process Compressors - the Influence of the Vaneless Diffusor Shape", Transactions of the ASME, Journal of Eng. for Power, Vol. 105, No.4, 1983, pp. 902-909 .

(10)Senoo, Y., Hayami, H., and Ueki, H., 1983, "Low-Solidity Tandem Cascade Diffusers for Wide-Flow-Range Centrifugal Blowers", ASME Paper 83-GT-3, 1983.

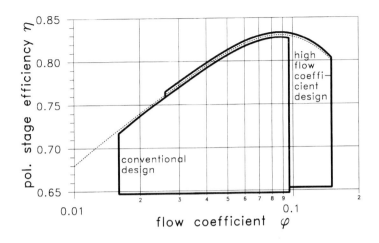

**Figure 1. Compressor efficiency increase
through high flow coefficient impellers**

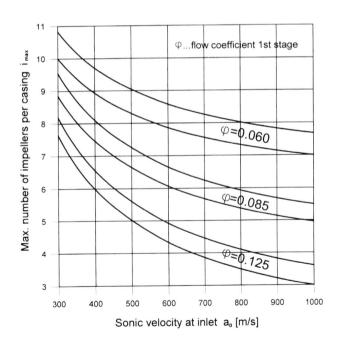

Figure 2. Maximum number of impellers per casing

C556/014 © IMechE 1999

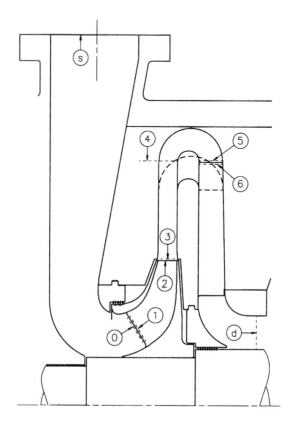

**Figure 3. Compressor stage
notation**

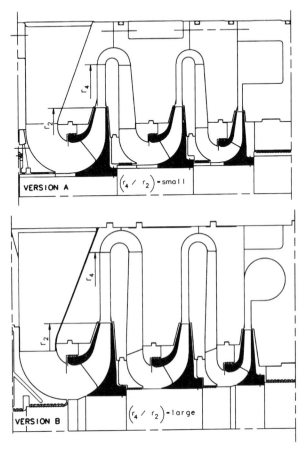

Figure 4. Two compressor sections with different diffuser
ratios and medium flow coefficient impellers

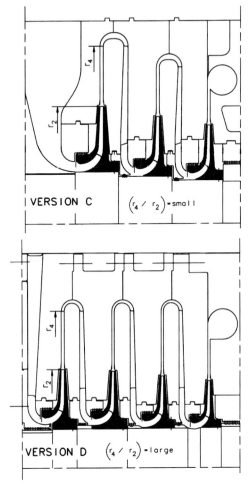

**Figure 5. Two compressor sections
with different diffuser ratios
and low flow coefficient impellers**

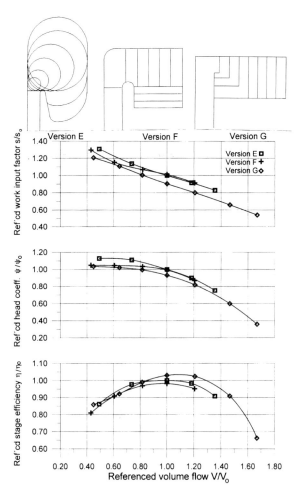

Figure 6. Three singlestage compressors with different volutes Versions E, F, G

C556/014 © IMechE 1999

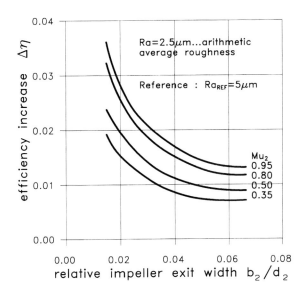

Figure 7. Increase of stage efficiency by reducing flow channel surface roughness

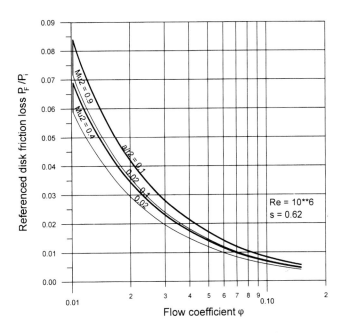

Figure 8. Impeller disk friction loss as a function of flow coefficient and axial clearance impeller / diaphragm

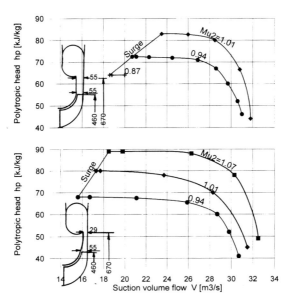

Figure 9. Singlestage compressor operating range, parallel walled versus highly tapered diffuser

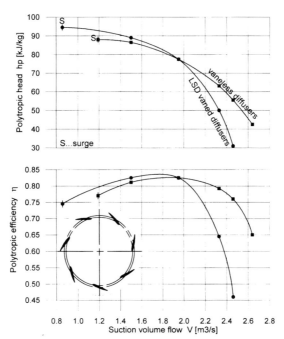

Figure 10. Six-stage compressor operating range, parallel walled vaneless diffusers versus diffusers with LSD vanes

C556/010/99

A new approach for reciprocating compressor wearband thickness detection

L G M KOOP
Thomassen International bv, Rheden, The Netherlands

SYNOPSIS

Reciprocating compressors are critical components in the process, gas and oil industries. Larger models are almost universally of a horizontal balanced opposed configuration.

The pistons require so called wear bands which function as bearings in supporting the weight of both piston and piston rod. Low wear is achieved by a combination of a suitable band material and lubricant. The trend for non-lubricated service, usually reduces band life substantially.

Due to the critical role of compressors in operationally expensive processes and the increasing importance of longlivety and reliability, knowledge of wear band condition, in particular is very desirable. Until now measurements were directed at establishing wear band thickness as a function of vertical displacement of the piston rod, measured at a location outside the cylinder. Transverse vibrations of the piston rod, crosshead clearance, load reversal and compressor operating conditions are unwelcome complications. Here the measured parameter is a derivative of the one of real interest: the true radial clearance between the piston and the liner.

Thomassen Compression Systems has initiated a development project for a reliable measuring method. A sensor has been developed, located directly in the cylinder wall, which measures accurately and reliably the true radial position of the piston relative to the liner. The sensor is stable and reliable under adverse operating conditions, with an accuracy of 10 microns or better. It is therefore highly suited for incorporation into condition monitoring systems.

CONTENTS

INTRODUCTION

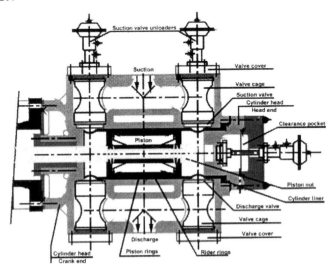

Cylinder

Reciprocating compressors are widely used in a great number of industrial applications. The main advantage of this compressor type is the low sensitivity for fluctuations of process conditions, and the large range of possible flow control. Only a few parts require regular maintenance and thereby limit the time between overhauls of the compressor. The variability of performance of the wear bands makes it necessary to monitor its performance. Nowadays the trend is to monitor the condition of equipment in order to perform condition based overhaul instead of time based overhaul.

Wear band monitoring of reciprocating compressors.

Monitoring of the wear of wearbands is usually done by measuring the vertical drop of the piston rod, several systems and suppliers are on the market. All these systems determine the wear rate using one or another derivative from the rod drop. Safety factors are built in to prevent the piston from running in the liner. On the other hand this means that overhaul is often premature. Once cylinders have been opened, the wear band will always be replaced whether they are fully worn out or not. Field experience indicates that the commonly used systems are not very reliable in their indications.

Discussion on rod drop measurements.

Rod drop measurements in order to determine the wear band condition is a derivative of the real radial piston position.

The measured vertical displacement of the piston rod is composed of the following components:
— Wear of the wear bands
— Thermal expansion of the piston relative to the cylinder
— Clearance of the crosshead
— Deformation of the piston during the stroke
— Lateral vibrations of the rod.
— Misalignment
— Wear of the rod in the stuffing box running section.

Based on the above considerations and on negative experiences in the field, our company decided to start a feasibility study for the design of a sensor which would directly measure the radial position of the piston.

SENSOR DESIGN REQUIREMENTS

Goal

A sensor should be designed to measure the radial position of the piston relative to the cylinder wall, and in this way establish the degree of wear of the rider rings

Location of the sensors

The sensors should be located inside the cylinder with the sensor head flush with the cylinder wall. As a result it will be directly exposed to the gas, and therefore subjected to the gas pressures and temperatures, which vary over a wide range.

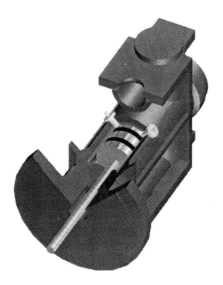

Fig. 2 Sensor location

Environmental Description

The design is for 250 bar pressure. The tip of the sensor can be subjected to 150 °C..

The sensor is in direct contact with the process gas, therefore it has to fullfill the possible requirements with regards to the corrosivity and safety.

The section of the casing where the electronic components are located is sometimes cooled by the cooling water. The maximum cooling water temperature is 90°C.

System Specification:

The sensor is designed to measure the distance between two objects, the following key features are important:

1.   Physically

Distance to be measured: 0.5 to 6.0 mm

Absolute distance measurement, contactless

Accuracy better then 100 µm (full range), stable for 2 years without calibration

Desired resolution 10µm

Environmental pressure sensor 250 bar

Environmental temperature sensor 150 °C

Vibration 2.5g 10Hz

Chemical resistance H_2S, CH's, acid gas

Insensitive to pollution, moisture, oil, PTFE and swarf of wear debris.

Possibility to replace the electronics without removing the sensor from the cylinder.

2. Electronic

Measuring cycle time <=1 msec, continuous measurement

2 distances should be measured simultaneously (2 sensor heads)

2 Sensors should be connected to one processing unit

Low energy radiation value

Certification according to Eex-IIC ia T4

Certification according to CE (immunity, emission)

Output signals:

Position information digital (serial, half duplex) through RS-485 cable (STP)

Maximum cable length: 1200 meter

Maximum bitrate 9600

Communication: galvanically separated

SENSOR TYPES

A great number of relative displacement sensors are available on the market. Below is a summary of sensors which are sometimes used in this field of application and some of the salient features:

a. Eddy Current
– Sensitive;accuracy to within 1 micron.
– Small
– Highly resistant to agressive environment
– Relatively insensitive for non-metallic contamination.

– Fixed cable length between sensor and driver
– Each sensor has to be calibrated for the electrical properties of the counter material. (type of material, hardening, coatings,etc.)
– The sensor has to be mounted in a chamber such that no distortion from the surrounding material will take place. For mounting inside a cylinder liner such a chamber would seriously influence the performance of the piston rings. Flush mounted shielded types are available. Due to the higher energy output they cannot be Certified for hazardous area's.
– High pressure and temperature applications are not available.

b. Optical
Optical sensors can be very accurate. When optical fibers are used as a sensor tip, they are small. The disadvantage is that inside a cylinder they would be sensitive to contamination, or mechanical distortion of the screen. Triangular methods whereby high accuracies can be reached, limit the range in which can be measured.

c. Acoustical
These sensors measure distance on the basis of the delay time of a sound wave between sensor head and object. This delay time is a function the velocity of sound, and is influenced by local vortexes and temperatures. The accuracy is 0.5mm maximum and therefor not suitable for our purpose.

e. Radar based
Accuray approximately 5mm

f. Inductive
Mainly used for proximity switches. The sensitivity is highly dependent on the counter material.

g. Capacitive
Available sensors on the market using this method are used for switches. Non contact displacement sensors are available. The accuracy can be very high. Eex certified sensors are not available.

	Eddy Current	Optical	Acoustical	Radar Based	Inductive	Capacity
Size	+	+	-	-	+	-
Accuracy	+	+	-	-	-	+
Contamination Sensitivity	+/-	--	+		+/-	+/-
Material influence	+	+	+		-	+
High Pressure Resistance	-	+	+		-	+
Aggressive Gases	+	+	+		-	+
Eex certification	+	-	+		+	-
Installation Requirements	-	+	+	-	+	-

HISTORY

1996:Feasibility study

The outcome of the feasibility study performed in cooperation with the electronic faculty of a technical College and an electronic development company in the Netherlands was:

- It is feasible to develop a dedicated sensor for this application.
- The sensing method which proved to be the most succesful was the capacity detecting method.
- A special electronic detection method needed to be developed in order to meet the Eex Certification requirements.

1997:Technical specification on mechanical properties

1997 FUSE application

FUSE is a project of the EEC that stimulates the application of electronics and electronic know-how in industrial enterprises. Thomassen has succesfully made an application for this project.

FUSE development phases
- Development and testing of 1st prototype
- Electronic redesign
- Redesign Sensor (smaller)
- Mechanical and electronic design and analysis
- GO/NOGO decision
- Fabrication of prototypes
- Static testing
- DynamicTesting
- End of FUSE 1998

METHOD

The sensor's working principle is to measure the change in electrical capacity between the sensor tip and the surface of the piston. When the rider rings wear, the piston will move radially downwards and therefore will be further from the sensor's head located in the cylinder wall. This displacement must be detected within an accuracy of 10 μm. The capacitive change is very small. This is caused by the fact that the diameter of the sensor must be as small as possible and the distance between the sensor and piston is relatively large.

The electrical capacity is a linear function of distance, the area of the capacitor plates, and the dielectric constants.

$$Capacity = \varepsilon_0 \cdot \varepsilon_r \cdot \frac{A}{d}$$

ε_0 = Diëlectrical Constant $8.85 * 10^{-12}$ [F/m]
ε_r = Relative Diëlectrical Constant
A = Area [m²]
d = distance between electrodes [m]

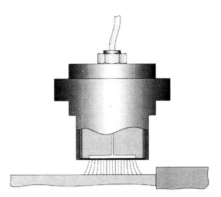

Fig. 3 Schematic sensor working principle

The linearity is distorted by (constant) parasitic capacities which are inevitably present in the sensor itself.

We have minimized the parasitic capacity by using special electronic components and optimizing the construction of the sensor. Further, the design features a guard system around the electrode inside the sensor. This guard is isolated from the casing, and has approximately the same potential as the electrode.

To minimize the capacity of the cable between oscillator and electrode the oscillator is incorporated in the casing of the sensor.

The relative diëlectrical constant depends on the medium between the plates. From literature studies it appears that for gases the ε_r deviates less than one percent from that of dry air. Liquids have a great impact on the ε_r. However the maximum layer thickness of liquid is limited, and has no major influence on accuracy.

The sensor is driven by a processing unit which is able to serve 2 sensors. It is envisaged that each cylinder will have its own processing unit. This processing unit is connected to a monitoring system via a RS485 digital connection.

Construction

The casing can be made of Cast Iron, or Stainless Steel. The tip surface material a ceramic. Also the isolator section is made from a ceramic material. This material is a good isolator and can withstand high surface pressures without showing extreme deformations. A further advantage of ceramics is that can be cemented with high strength adhesive..

The separate parts are assembled with a special type of adhesive. The casing with the ceramic inserts has been designed using Finite Element Calculations.

The electrical contact between the board, the electrode and guard used goldplated pins. The wiring between the sensor and the processing unit is led through watertight Stainless Steel tubing.

The sensor is mounted in the cylinder wall using a bolted flange (not threaded) and the sealing against the gas and cooling water is obtained by using O-ring type seals. The fine adjustment of the radial position is obtained using shims.

The sensor tip has the same curvature as the cylinder liner. In the mounted postion it has some circumferential clearance to avoid problems when the liner expands. This gap is closed by an elastic seal type. This is neccessary because otherwise short-circuit flow will occur across the piston rings when they pass the sensor.

System description:

The system is divided in two parts.
First part is the actual sensor, which produces a frequency that has a direct relationship with the distance between sensor and object. The distance measurement is based on measuring the change in capacitance between two plates. The capacitance is fed into an active electronic circuit, where it is translated to a frequency. The frequency is related to the measured capacitance.
Because the distance has to be measured at two locations at the same time, there are two sensors in each system. Both frequencies are connected to one processing unit.

Second part is the processing unit. Here the in- and outgoing signals are isolated, in order to fulfill the Ex certification demand. Both frequencies are counted using a Complex Programmable Logic Device (CPLD). At this point both frequencies are available in a digital form and can be read by the microcontroller. However the measured frequency does not have a linear relation with the distance. For calibration, the system may be switched to a calibration capacitor and measure the frequency. This gives information in static situations, assuming the reference capacitor has a constant value. Any variations caused by temperature drift etc. can be corrected this way. Since this has to be realized in real time, Digital Signal Processing (DSP) techniques are used, implemented using VHDL (Programming Language) in the CPLD chip.
The microcontroller is used to send the information to the monitoring system (serial), and to support initial calibration functions. It also provides system information for service purposes (led-indicators to monitor system functions).

TESTING

Pressure and leak testing
The sensors have been tested on 375 barg hydrostatic test pressure. A leaktest has been performed with Helium at 250 barg. During these tests the electronic circuits were active and a measurement of a fixed distance from a steel plate was performed. After testing the adhesive and the material showed no damages or leaks. The electronic circuits remained intact.

Static Calibration

The static testing of the sensors was performed on a full scale test rig, on which the curvature of the piston is simulated by a pipe section. This rig will also be used for the calibration of every new sensor. Figure 4 shows the measured output against the actual distance of the sensor tip from the piston surface. We have calibrated the sensor under the following conditions:

1. with a clean head opposite to the steel surface of the piston
2. with a clead head opposite to the PTFE wear band material
3. with a head contaminated with grease
4. with a head contaminated with a solution of grease and PTFE.

It appeared that the slope of the curve did not deviate to a great extent in comparison with a non conducting contamination on the head. The offset of the curve is caused by the layer thickness of the contamination. In this case the PTFE showed a almost identical characteristic as Steel or Cast Iron.

Conducting materials with a considerable layer thickness disturb the measurements considerably.

Dynamic testing

For dynamic testing, a full scale refinery compressor is available.

This compressor is a 2-stage air compressor with cylinder bore's of 540 mm and 340 mm. The pressures, the loading conditions and the speed can be varied. The temperature of the air to the suction 2nd stage can be controlled.

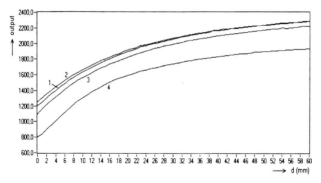

Fig.4 Calibration curves

On our test compressor we have performed series of measurements to test the stability of the sensor and electronics. In the 2nd stage cylinder we have installed two sensors. The sensors operated in a corrosive environment of hot and relative damp air. By cooling down the inlet temperature of the 2nd stage, wet air entered the cylinder. In this way a very corrosive condition can be created.

During the stroke the sensor position remains between the piston rings, and therefore only registers the passing of the cast-iron of the piston and the PTFE of the wear bands. The wear bands are of the shrink fit type.

In fig.5. the diagram of the a measuring cycle is shown. The flat top of the curve represents the position of the piston, the flat bottom the postion on the wear bands. This figure has been corrected for the non-linearity.

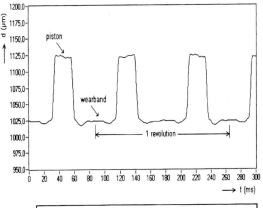

Fig. 5 Plot of measuring cycle of 1 sensor

The exact position of the piston relative from cylinder wall can be establised by calculating the vector sum of both sensor readings. The zero value is the offset when starting the measurement cycle. Fig 6. shows the position of the topsurface of the piston.

The graph shows measurement results of a week continuous running, with a stop interval during the weekend.

After the running test the compressor cylinder has been disassembled . The casing of the sensor was heavily corroded and the surface of the ceramic tip contaminated with PTFE and dirt particals.

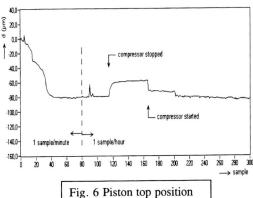

Fig. 6 Piston top position

C556/010 © IMechE 1999

Conclusions from measurements

– It is possible to design and built a sensor which can directly measure the piston displacement.
– The carbon filled PTFE is detected just as well as the cast iron.
– The flexibility of the developed software makes it possible to adapt the sensor position and sensitivity for almost all types of cylinders.
– The sensor is very accurate tool. The slight changes of the cooling water temperature could be read in a slight change of piston position.(see fig. 6 after sample 200)
– To avoid contamination of dirt and oil particles the sensors must be located in the top half section of the cylinder. The piston rings will wipe the sensor head every time they pass.
– From measurements in the test compressor it appeared that during the stroke the piston does not move exactly parallel with the cylinder axis. The rider band will accordingly not wear only in the vertical axis. Two sensors are necessary to measure the exact location of the piston relative to the cylinder wall. A single sensor could be employed on the vertical axis or a near vertical axis, if a lower grade of accuracy is acceptable.

C556/031/99

Thermodynamic–conventional tests on two 4 MW pumps

M A YATES
AEMS Limited, Ottery St Mary, UK
A KUMAR
Sulzer (UK) Pumps Limited, UK

Synopsis

The paper details both the thermodynamic and conventional tests undertaken on two 4MW crude export pumps. The pumps were of the horizontal barrel casing type, with eight stages. The pumps were driven by an induction motor, through a fluid coupling. The pumps ran at approximately 3500 rpm. To achieve increased discharge pressure and output from the pumps, new cartridges were fitted. The cartridges incorporated new hydraulics, with are described in the paper. To verify the claimed improvement in performance, thermodynamic tests were undertaken independently at the manufacturers works and compared with the conventional test results.

NOMENCLATURE

η_P	Pump Efficiency
η_{PI}	Internal Pump Efficiency
η_M	Motor Efficiency
η_G	Gearbox Efficiency
η_D	Drive Efficiency
m^3/hr	Cubic metres per hour
BPD	Barrels per day.
W	Work Done
Q	Volume Flow Rate
H	Head
DT	Change in Temperature
Cp	Specific Heat Capacity of Fluid

P_{gr}   Motor Input Power
G   Gravitational Acceleration
ρ   Density of Fluid

1 INTRODUCTION

AEMS Ltd were commissioned by Norsk Hydro, to undertake a comparison Yatesmeter thermo- dynamic pump test in conjunction with a conventional pumps test, been performed by Sulzer (UK) Pumps. The tests were undertaken on two new cartridges for the Troll platform, in the Norwegian sector of the North Sea. The pumps' function is to export crude oil, and the new cartridges were part of a plant upgrade scheme.

The tests were undertaken at the manufacturers works in Leeds (United Kingdom) using cold water. However, due to the significant heat generated within the test loop, cooling was required to maintain the temperature within the limits of the ISO standard. Cooling was provided by the use of a cooling tower and fan.

The units were tested through out the operating range, both using conventional and Yatesmeter (Thermodynamic) testing techniques. The upgrade of the pumps enabled an increase in production, from 240,000 BPD to 280,000 BPD.

Both efficiency and performance of the pump was paramount importance to Norsk Hydro, and in view of this, they required the original equipment manufacturer, Sulzer to undertake extensive tests on the new cartridges.

To fulfil Norsk Hydro's requirement, the following tests were undertaken:

Quantity Vs Pump Head
Quantity Vs Pump Efficiency
Quantity Vs Shaft Power
Vibration measurements on all bearings.

2 PUMP HYDRAULIC DESIGN

In reviewing the hydraulic design, it is helpful to compare the original and upgraded duty requirements.

	Original	Upgraded	Remarks
Flow m3/h	534	610	14% Increase
Head m	1773	1867	5% Increase
Speed rpm	3473	3465	
Efficiency %	80	80	
Power kW	2880	3440	19% Increase

The constraints imposed on the upgrade were

a) Existing motor was to be retained. Therefore the maximum available motor power (including fluid coupling losses) placed a limit on permissible increase in pump output.

b) Pump casing dimensional limits. Existing barrel casing and ring sections were to be retained.

c) Pump casing pressure limitations. This meant that the shut off head could not increase above the original value.

Constraint (a) was removed when the motor manufacturer confirmed that with only minor modifications, the motor rating could be increased to accommodate the upgraded duty.

Constraints (b) and (c) implied that the increased head and flow had to achieved with no increase in impeller diameter. In other words, the new impeller and diffuser had to be designed for a much higher head coefficient than the existing design (from 0.98 to 1.08).

Practical design correlations indicated that higher head coefficients could be achieved by increasing the impeller outlet width and angle with corresponding changes in the diffuses length and throat area. Confirmation that the increase in head coefficient would indeed be achieved by the above modifications was sought by calculating the flow in the impeller/diffuser passage using a commercially available 3D Navier Stokes program (TASC Flow). This showed that the predicted head would be achieved with a tolerance which was as good as the test measurements. Details of the original and new hydraulic design are tabled.

Parameters	Original	New
Impeller Diameter mm	355	350
Outlet width mm	21.5	29.0
Number of vanes	7	7
Outlet angle deg	25	30
Diffuser throat area sq.mm	6050	6890

3 TEST RIG DETAILS

The comparison tests were undertaken at the manufacturers works in Leeds (United Kingdom).

The test loop consisted of a closed loop circuit and is shown in Fig 1. The loop incorporated a booster pump, which was used to prime the 8 stage horizontal barrel pump under test. The pump was driven by an 8MW slave motor and fluid coupling. The water was cooled by a cooling tower. The following instrumentation was used on the test rig.

(i) Thermodynamic Test

   Yatesmeter Pump test equipment

(ii) Conventional Test

   Delivery Electromagnetic Flowmeter
   Balance Water Electromagnetic Flowmeter
   Suction Pressure Transducer
   Delivery Pressure Transducer
   Bourden Gauges for Display
   2 Wattmeter Powermeter

See enclosed diagram for schematic of test loop.

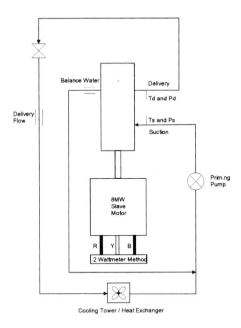

Cooling Tower / Heat Exchanger

4 TESTING METHODS

Two methods of pump testing were used.

Yatesmeter (Thermodynamic)
Sulzer Test Rig (Conventional)

4.1 Yatesmeter (Thermodynamic) Testing

Fundamentally, the Yatesmeter measures the rise in differential temperature across the pump in relationship to the generated pump head. The relationship between the two properties enables the pump efficiency to be assessed, without the requirement to measure either the quantity pumped or shaft power to the pump. With the addition of power to the standard pump equation, quantity can be calculated. The configuration shown in Fig 2 and 3.

Parameters measured / used by the Yatesmeter for the calculation of the pump parameters:
    Differential Temperature
    Suction Pressure
    Delivery Pressure
    Power to Motor
    Motor Efficiency
    Gearbox Efficiency

Calculated pump parameters include:
  Internal Pump Efficiency
  Pump Head
  Suction Head
  Delivery Head
  Pump Flow rate
  Power to Motor
  Power to Pump

Because of the particular test rig configuration, with the use of a slave motor, the Power to Motor, Motor Efficiency and Gearbox Efficiency was measured by Sulzer, and as a result the data is common in both sets of calculations. In addition, the balance water flow was removed from the delivery manifold before the delivery temperature probe. As a result the Yatesmeter measured the Internal Pump Efficiency. To convert the Internal Pump Efficiency into actual Pump Efficiency the ratio of the balance water flow to the delivery manifold flow was used to correct the Yatesmeter's Internal Pump Efficiency calculation into actual Pump Efficiency.

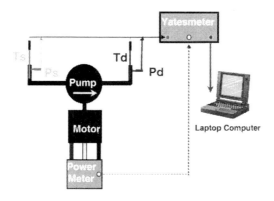

FIG 2: Yatesmeter Layout Diagram

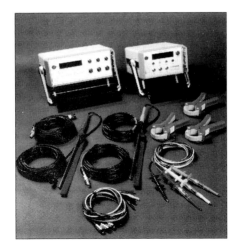

FIG 3: Yatesmeter Equipment

-4.2 Yatesmeter Theory

A simplified form of the general calculations are given as follows:

Pump Efficiency, $\eta_p = \dfrac{W_{out}}{W_{in}}$

but $W_{in} = W_{out} + losses$

Thus, $\eta_p \quad = \dfrac{1}{1 + \dfrac{losses}{W_{out}}}$

But, $\dfrac{losses}{W_{out}} \quad = \dfrac{dT \ Cp \ Q}{\rho g Q H}$

Therefore, $\eta_p \quad = \dfrac{1}{1 + \dfrac{dT \ Cp}{\rho g H}}$

Volume Flow Rate $Q = \dfrac{P_{gr} \eta_m \eta_p}{\rho g H}$

But, $\eta_p \quad = \dfrac{1}{1 + \dfrac{dT \ Cp}{g H}}$

$$\text{Therefore, Q} = \frac{P_{gr}\eta_m}{\rho gH + (dT \quad Cp)}$$

4.3 Conventional Testing

With conventional testing, pump efficiency is calculated from the measurement of the following parameters:

Pump Head
Pump Flow Rate
Power to Pump

To measure the pump head, the test rig used two pressure transducers, with Bourden gauges to display the value.

The pump flow was measured by two electromagnetic flowmeters. The first flowmeter recorded the flow on the delivery manifold after the balance water tee, while the second flowmeter recorded the balance water flow.

The Power to Motor was measured by the two wattmeter method.

Both, the motor efficiency and gearbox efficiency, for each test point was calculated by a Sulzer computer program.

Conventional formula

Pump Efficiency equals

$$\eta_P = \frac{\rho gQH}{P_{gr}\eta_M}$$

5 PUMP TEST RESULTS

The following section details both the individual and comparison results of both methods of pump testing.

5.1 Yatesmeter (Thermodynamic) Results

The following tabulated data is for the thermodynamic test on cartridge No.1

Primary Data

The primary data was the parameters measured by the Yatesmeter and the powermeter. This information was used for the calculation of pump efficiency.

DATA SET 1

Yatesmeter Primary Data

Differential Temp 'mK	Water Temp `C	Measured Suction Pressure m	Measured Delivery Pressure m	Motor Eff %	Gearbox Eff %	O/A Drive Eff %	Power to Drive kW	Shaft Power kW
1,243.92	29.01	48.84	1,885.55	93.46	98.2	91.79	4,177.00	3,833.95
1,120.69	30.30	48.22	1,793.58	93.36	98.2	91.72	4,384.00	4,020.87
1,093.51	32.24	47.29	1,594.04	93.64	98.3	92.05	4,707.00	4,332.71
1,170.44	33.42	53.44	1,453.65	93.83	98.3	92.25	4,868.00	4,490.91
1,777.90	28.94	47.59	1,980.10	93.61	98.1	91.87	3,765.00	3,458.86
2,517.23	27.10	48.09	2,054.11	93.47	98.1	91.69	3,594.00	3,295.48

Primary Calculated Data

The primary calculated data table contains the calculated results for the performance of the pump during the test.

DATA SET 2

Yatesmeter Calculated Data

Pump Head m	Motor Eff %	Gearbox Eff %	Drive Eff %	Power to Drive kW	Shaft Power kW	Pump Speed rpm
1,837.70	93.46	98.2	91.79	4,177.00	3,833.95	3,456
1,746.60	93.36	98.2	91.72	4,384.00	4,020.87	3,450
1,548.50	93.64	98.3	92.05	4,707.00	4,332.71	3,453
1,402.30	93.83	98.3	92.25	4,868.00	4,490.91	3,459
1,933.10	93.61	98.1	91.87	3,765.00	3,458.86	3,473
2,006.40	93.47	98.1	91.19	3,594.00	3,295.48	3,477

Internal Pump Flow m³/h	Balance Water M³/h	Pump Flow m³/h	Internal Pump Eff %	Pump Eff %
641.89	18.2	623.69	83.8	81.43
719.31	17.9	701.41	85.1	82.99
857.85	16.9	840.95	83.5	81.87
941.38	16	925.38	80.1	78.72
506.9	18.5	488.4	77.2	74.36
415.7	18.9	396.8	69.3	66.17

Site Corrected Data

The site corrected calculations contain the pump performance results once they have been corrected for both the site conditions and the fluid conditions.

DATA SET 3

Yatesmeter Data Speed Corrected to 3465 rpm and Corrected for Average Product Specific Gravity.

Pump Head m	Pump Flow m^3/h	Pump Eff %	Shaft Power kW (water)	Shaft Power kW (crude oil)
1849.1	621.3	81.4	3842.7	3420.0
1761.8	704.5	83.0	4073.5	3625.5
1557.7	844.7	81.9	4378.0	3896.5
1405.8	927.9	78.7	4514.3	4017.7
1926.1	486.8	74.4	3435.0	3057.2
1996.6	394.6	66.2	3261.5	2902.7

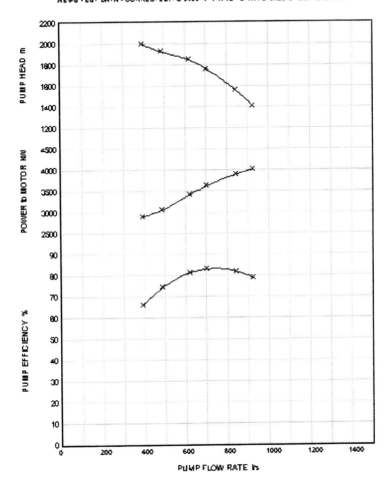

8 STAGE HORIZONTAL BARREL CASING PUMP
Pump: CARTRIDGE No.1   Tester: MAY/AMY   Test date: 19th JUNE 1997
AEMS TEST DATA - CORRECTED: TO 3465 RPM AND FOR AVERAGE SPECIFIC GRAVITY

FIG 4 YATESMETER RESULTS –SITE CORRECTED

Yatesmeter Time Data
The time based traces show how the differential temperature and water temperature vary through out the duration of the pump test.

Yatesmeter Differential Temperature Time Trace

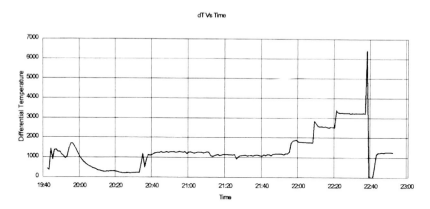

dT Vs Time

Note
The trace shows both the initial change in dT as the unit was prepared for testing (19:40 to 20:40) and the effects on dT due to the change in pump efficiency, through throttling.#

FIG 5 DATA SET 5

Yatesmeter Water Temperature Time Trace

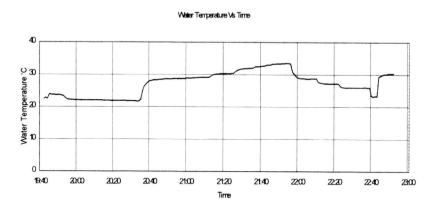

Water Temperature Vs Time

Note
This trace shows how the water temperature was controlled by the use of the cooling tower to maintain the temperature within the limit of the ISO standard.

FIG 6 DATA SET 6

5.2 Sulzer (Conventional) Results
The following tabulated data is for the conventional test on the same cartridge (No.1)

Sulzer Results
The tabulation of the Sulzer test results corrected to site condition.

DATA SET 7

Sulzer (Conventional) Test

The following tabulation contains the site corrected data from the conventional test under taken by Sulzer.

Sulzer Data Speed Corrected to 3465 rpm and
Corrected for Product Specific Gravity. – Data 7

Pump Head m	Pump Flow m^3/h	Pump Eff %	Shaft Power kW
2133.1	142.2	32.9	2232.6
2069.2	340.3	59.1	2882.2
2010.3	399.3	68.9	2903.8
1942.0	496.6	76.3	3058.9
1869.7	608.7	79.9	3443.4
1865.8	615.8	80.8	3437.5
1772.9	705.8	83.3	3631.7
1576.0	840.0	82.1	3902.4
1425.2	936.1	80.2	4025.3

Sulzer Graphs

The Sulzer graph contains the information of the site corrected data. The information has been displayed in the typical 3 y axis to 1 x axis.

C556/031/99

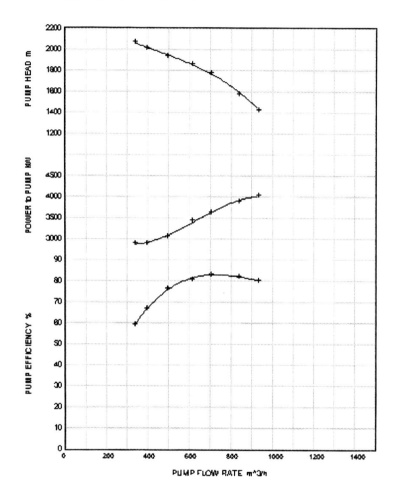

FIG 7 SULZER (CONVENTIONAL) RESULTS– DATA SET 8

6 DISCUSSION ON RESULTS

The following tabulation was produced to show the difference between the two methods of testing. For each of the calculated parameters of Pump Head, Pump Efficiency and Pump Flow the difference between the results has been examined. The difference calculation has been calculated at the closest test point to the duty condition of the pump. Because the power measurement was common for both, this has been excluded from the analysis.

Specified Pump Duty Condition

Pump Head	1867.0 m
Pump Flow	610.0 m³/hr
Shaft Power	3444.0 kW
Pump Efficiency	80.0%

Comparison between Test Results

Yatesmeter Head	1849.1m
Sulzer Head	1865.8m
% Difference	0.9%
Yatesmeter Pump Efficiency	81.4%
Sulzer Pump Efficiency	80.8%
% Difference	0.7%
Yatesmeter Pump Flow	621.2m³/hr
Sulzer Pump Flow	615.8m³/hr
% Difference	0.9%

The overall variation between the two test methods was less than 1%.

The following graph shows the difference between the two test methods when compared with each other.

A comparison graph is shown in Fig 8

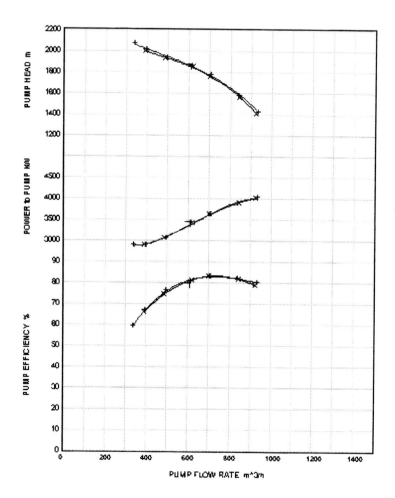

8 STAGE HORIZONTAL BARREL CASING PUMP
Pump: CARTRIDGE No.1   Tester: WAY/AWY   Test date: 19th June 1997
AEWS TEST DATA - CORRECTED: TO 3465 RPM AND FOR AVERAGE SPECIFIC GRAVITY ✕
MANUFACTURERS WORKS TEST +   DUTY POINTS ⌐

FIG 8: COMPARISON OF RESULTS – DATA SET 9

7 CONCLUSION

The thermodynamic method showed an excellent comparison with the conventional test. The discrepancies between the two tests were well within the pump test standards. One would expect the thermodynamic test results to give a greater efficiency as it is measuring the true hydraulic performance of the pump. The method excludes the radiated bearing loss.

The thermodynamic method can accommodate large fluctuations in suction temperature, without effecting the uncertainty of the pump efficiency measurement.

The most important facet of the thermodynamic test is that it is a transferable standard. This means that the test can be replicated on the platform and the performance of the pump can be assessed in its operating environment.

C556/031/99

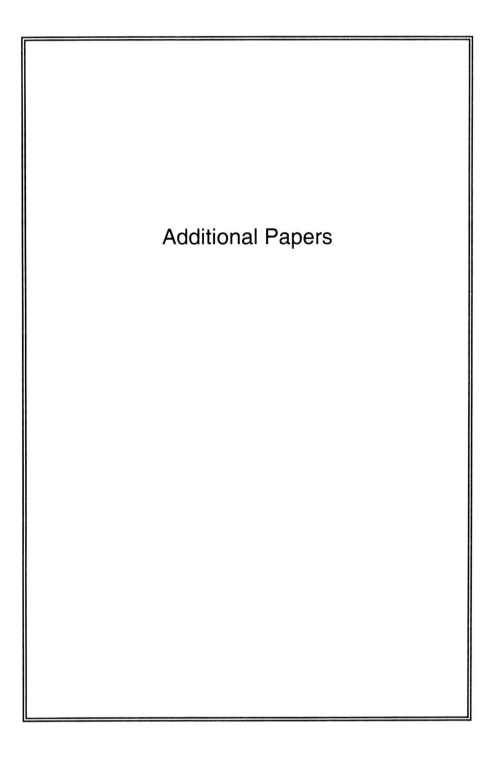

Additional Papers

C556/004/99

Simulation of transients of high amplitude in pipe systems

J M BOERSMA and **K N H LOOIJMANS**
TNO Institute of Applied Physics, Delft, The Netherlands

SYNOPSIS

Fast high-amplitude transients ask for a non-linear modelling approach in which large density variations and heat exchange can be considered. Operation of safety-valves, relief valves, the occurrence of valve failure and the start-up or shutdown of rotating equipment in industrial pipe systems can lead to the emergence of such high-amplitude transients. The subsequent propagation of the high-amplitude waves in the pipe system can cause severe damage to the piping and process equipment. Recently, a numerical code called 'PULTRAN has been developed that enables the prediction of high-amplitude transients and their effect on pipe systems and rotating equipment. The basics and possibilities of PULTRAN are described in the paper.

1. INTRODUCTION

Modelling of flow dynamic phenomena has been a research topic at the Institute of Applied Physics of TNO since 1968. This has already resulted in a digital simulator called PULSIM (PULsation SIMulator), which is used for the investigation of pulsations in pipe systems. In pipe systems a variety of flow dynamic phenomena can occur. Modelling of these phenomena is important for the design and operation of pipe systems. Though the physics of these phenomena is the same, there is a considerable difference in applications, each of which require a different approach.

The PULSIM code developed at TNO-TPD for performing pulsation studies, is not suitable for the prediction of high-amplitude pressure waves in compressible fluids due to the method used which is based upon linear characteristics. Operation of safety-valves, relief valves, and the occurrence of valve failure in industrial pipe systems for gas handling and transport can lead to the emergence of such high-amplitude transients. The subsequent propagation of the high-amplitude waves in the pipe systems can cause severe damage to the piping and process equipment and accurate prediction of these waves demands advanced numerical techniques. Recently, a numerical code has been developed that enables the prediction of high-amplitude transients and their effect on pipe systems. Effects of wall friction, heat conduction, and non-ideal gas behaviour are integrated in the new code which is called 'PULTRAN'. In this paper, the solution method and the possibilities of PULTRAN are discussed.

2. NUMERICAL METHOD

The numerical method used for solving the describing equations in PULTRAN is called the Random Choice Method (RCM). It is a relatively new numerical method, first described by A.J. Chorin (1) and has especially proven to be successful in the description of quasi-1-dimensional unsteady compressible flow with pressure waves of high-amplitude. The TNO-TPD version of the RCM-code is similar to the one described by Toro (2). The RCM solver deals with non-linear effects by using a dedicated numerical method. Discontinuities like shock waves and contact surfaces are automatically handled. Salient feature of this method is the absence of numerical diffusion and dispersion. Therefore, the exact description of shock waves and contact surfaces does not require special measures. Some other virtues of the RCM are (2) :

- ability to handle wave interactions efficiently and automatically.
- absence of adhoc procedures and tuning of arbitrary parameters
- good physical and mathematical foundations

One limitation of the RCM was the needed computational time to solve the Riemann problem exactly, but this has been overcome by Gottlieb (3) who has presented a new exact Riemann solver. In his method, the Riemann problem is reduced to a single non-linear algebraic equation (instead of three as in the Godunov's solution). This results in a highly efficient method; the required calculation time is limited.

The Random Choice Method is applicable to quasi-linear hyperbolic systems of conservation laws

$$\frac{\partial U}{\partial t} + \frac{\partial F(U)}{\partial x} = C \quad \text{or} \quad \frac{\partial U}{\partial t} + B.\frac{\partial U}{\partial x} = C \qquad (1)$$

where $B = \dfrac{\partial F}{\partial U}$ represents the Jacobian matrix for F(U).

The system is denoted as hyperbolic if :
- all eigenvalues of B are real, and
- B can be converted to a diagonal matrix.

The solution procedure starts with an initial estimate U(t=0). The first step concerns the discretization of these initial data with respect to the spatial co-ordinate x. This is illustrated in figure 1. The data are approximated by a piecewise constant function of x.

The sequence of Riemann problems RP(i,i+1) at time t_n is defined as a set of initial value problems

$$U(t_n, x) = \begin{cases} U_i & if \ x \le (i + \frac{1}{2})\Delta x \\ U_{i+1} & if \ x \ge (i + \frac{1}{2})\Delta x \end{cases} \tag{4}$$

The solution to this Riemann problem, when represented in a x-t domain, looks like figure 2 and consists of a right, a left and a middle wave. Both right and left waves can be either shocks or rarefaction's and the middle wave is a contact discontinuity which separates gases originally in cell i from those originally in cell i+1.

Given ρ, u and P at states i and i+1, the solution of the Riemann problems is completely determined if the pressure p^* and velocity u^* for the region between the left and right waves are known. p^* and u^* are constant in this region, but the density is piece-wise with two values ρ^*_l and ρ^*_r on the left and right side of the contact discontinuity respectively. All the thermodynamic and dynamic properties can be calculated exactly from an algebraic consideration of the transition relations for shock fronts and rarefaction waves (4). These variables are calculated by means of an exact Riemann solver using Newton-Raphson iteration (2).

The RCM is a time explicit method; the solution at a time level t_{n+1}, $U(t_{n+1},x)$, is calculated by solving a sequence of Riemann problems at the foregoing time level t_n. Having solved the Riemann problem at time t_n, the exact solution at time t_{n+1} has a range of values that depend on the x-position. On the sampling interval [(i-1/2)Δx, (i+1/2)Δx)] the Random Choice Method selects an x-position at random and the corresponding values of the solution are assigned to the nearest computational grid node i (figure 3). The van der Corput sequence is applied for the random number generation (5), whereby an alternating sampling position left and right from the node i is selected. The sample area Δx then covers two Riemann problems.

The procedure described above is repeated for each time step. After a large number of time steps, the averaging property of the random selection of solutions of the Riemann problems will result in a correct description of the wave propagation in the x-t domain. The computational speed is determined by the time step Δt used which in turn is limited by the Courant-Friedrichs-Lewy (CFL) condition. The CFL condition simply says that the time step should be chosen small enough so as to avoid wave interaction of neighbouring Riemann problems before the random sampling of the interval is carried out.

The Random Choice Method has been extensively tested against measurements. Examples can be found in (6) and (7).

In PULTRAN the RCM is used to solve the following set of equations :

Mass : $\dfrac{\partial \rho}{\partial t} + u.\dfrac{\partial \rho}{\partial x} + \rho.\dfrac{\partial u}{\partial x} = 0$

Momentum $\dfrac{\partial u}{\partial t} + u.\dfrac{\partial u}{\partial x} + \dfrac{1}{\rho}.\dfrac{\partial p}{\partial x} = -\dfrac{1}{\rho}.f_w = -\dfrac{2f}{D}.u.|u|$ (2)

Energy : $\dfrac{\partial p}{\partial t} + \rho a^2.\dfrac{\partial u}{\partial x} + u.\dfrac{\partial p}{\partial x} = (\gamma - 1).\{Q + u.f_w\}$

where f_w is the wall friction, f the fanning friction factor, Q the heat exchange with the environment, a the velocity of sound, γ the ratio of specific heats and u, p, ρ the velocity, pressure and density respectively. This can be written in the form of equation (1) with

$$U = \begin{bmatrix} \rho \\ u \\ p \end{bmatrix} \quad and \quad B = \begin{bmatrix} u & \rho & 0 \\ 0 & u & 1/\rho \\ 0 & \rho a^2 & u \end{bmatrix} \quad\quad (3)$$

In PULTRAN several physical phenomena are considered like :

- wall friction
- heat exchange with the environment
- wave interaction with changes in pipe diameter
- possible choking e.g., in restrictions, valves, and at pipe expansions
- real gas effects

Wall friction causes damping of pressure waves, while heat exchange influences the propagation velocity of waves. Especially in long pipe systems, accurate calculation of wave propagation requires the inclusion of these effects in the calculation. In compressible flow, an important phenomenon known as choking can occur. If the pressure downstream of a nozzle is lowered progressively while the upstream pressure remains constant, the mass flow rate of the gas increases until it reaches a limiting maximum value at a critical downstream pressure. Reduction of the downstream pressure below this critical value does not result in any increase in mass flow rate and the nozzle is said to be choked. Often the occurrence of choking is desired, for example to achieve a large pressure drop in the system. However, choking can also limit the controllability in emergency situations and therefore it is important to know the locations in the system where choking can occur.

When dealing with gases at elevated pressures and/or very low temperatures real gas effects may be significant and alternative equations of state or a compressibility correction factor should be used. In PULTRAN gas properties are supplied as look-up tables which are functions of pressure and temperature. This offers a great flexibility and even vapours such as steam can be treated.

The set of equations (2) has been extended by a large number of boundary conditions like T-junctions, orifices, reducers, volumes, valves, pistons, closed ends and infinite pipes. In PULTRAN, an installation is modelled using pipe elements which are connected by these boundary conditions. Also boundary conditions are available for different types of fluid machinery. This way, the effect of pulsations in the pipe system on valve dynamics of reciprocating compressors or possible surge of centrifugal compressors can be predicted.

3. EXAMPLE

The selected example is a gas production system (see figure 4) consisting of the production well (270 bar), a pressure control valve (PCV) which reduces the pressure from 270 to 100 bar, a High Integrity Pipeline Protection valve (HIPP), a transport line and a liquid separator. The pipe system downstream a HIPP valve has a maximum working pressure of 150 bar. The gas temperature is 334 K. The temperature of the environment is 293 K. The gas produced contains a few mass percent liquid which can accumulate in the lower parts of the transport line. Thus a liquid slug can build up which can block the line when a sudden increase in the gas flow occurs. The HIPP valve is activated at 120 bar. This pressure can occur for instance when the pressure control valve fails and an open connection arises with the high pressure pipe system. The question to be answered by means of model calculations is what the maximum pressure downstream the HIPP valve can be if this valve shuts within 2 seconds after activation. As a worst case it is assumed that downstream the HIPP valve the line is blocked by a liquid slug which was built up there during earlier operations.

The model of this system can be simplified for the purpose of calculating the transient phenomena caused by break-out of the choke in the pressure control valve (see figure 5). Regarding the response time of the HIPP, a time interval of 5 seconds is considered after break out. In this time interval, no wave reflection from the well will occur and therefore the distance between well and orifice is not modelled. Instead an endless line boundary condition upstream of the orifice is considered. This boundary condition gives no wave reflection.

3.1. Results in the case of real gas behaviour, wall friction and heat transfer.
In steady state conditions (i.e. normal operation of PCV) a flow of 10 [kg/s] exists. The density upstream and downstream the PCV is 190.7 and 72.5 [kg/m^3] respectively, and no choking conditions occur in the installation.

These initial conditions are used as a start for the calculation of the break out of the choke of the pressure control valve at t = 0 [s]. At this time the total pressure loss coefficient over the PVC reduces to zero instantaneously.

Figures 6 to12 show the calculated change in pressure, density, velocity and temperature at the positions indicated in figure 5. The closing of the HIPP valve is not considered in the calculations. Figure 6 shows the conditions just after the orifice. After about 0.2 seconds, the upstream travelling expansion wave reaches the orifice and the pressure drops below 100 bar.

Due to the high pressure difference and the resulting high velocity, choking conditions occur at the orifice. Subsequently, the pressure recovers and increases fast due to the right travelling shock wave which was reflected at the liquid slug. The number of interacting right and left travelling waves becomes higher as time proceeds. Thereby the amplitude of the waves becomes lower due to friction and heat transfer to the surroundings while the temperature of the gas after the orifice decreases due to the expansion process. Figure 10 shows the conditions at the HIPP valve. The pressure at the HIPP valve increases to 140 bar in 0.1 seconds after break out of the control valve. So within 0.1 seconds the HIPP valve is activated. An expansion wave follows the shock wave and reaches the HIPP valve after 0.45 seconds. After that the pressure at the HIPP valve increases and would exceed the allowable level of 150 bar after 4.5 seconds. However, at that time the HIPP valve will be closed so this pressure will not be attained. Looking at the conditions at the HIPP valve everything seems acceptable. However, the pressure build-up at position 6 (figure 11) is much faster due to the positive reflection of the shock wave by the liquid slug. At this location, the allowable level of 150 bar is exceeded within 0.6 seconds. Therefore, the HIPP valve should close much faster or the pipe system has to be changed.

A possible solution is the installation of an orifice downstream the HIPP valve. Due to this orifice, the amplitude of the first shock wave travelling downstream will be lowered. The calculated results for an orifice with a bore of 200 mm at a position of 100 meters downstream the HIPP, are shown in figure 13. The static pressure drop over this orifice at normal operation is only 0.01% of the mean pressure and should therefore present no problem.

3.2. Results in the case of ideal gas behaviour and neglecting wall friction and heat transfer
Also shown in figures 6 to 12 are the calculated results for ideal gas behaviour, no wall friction and no heat exchange (denoted by dashed lines). Figure 6 shows clearly that this latter approach results in much higher predicted pressure fluctuations. If these calculated results are used for the design of the installation, the selected pipe wall thickness will be much too high. From a safety point of view this is not necessary. Using the PULTRAN approach of section 3.1 will result in a safe pipe system at lower costs.

3.3. Mechanical response study
The calculated dynamic forces are used as input for a subsequent mechanical transient calculation using finite element analysis (available programs CAESAR-II and ANSYS). This way, possible fatigue failure of the installation can be predicted at the design stage. Both geometrical linear and non-linear systems, e.g. pipe systems with 'gap' supports, can be handled. A special interface takes care of the transfer of the calculated transient forces from the acoustical PULTRAN model to the finite element model in an easy and efficient way. Results of the study are displacements, cyclic stresses, support reaction forces and moments which can be compared with allowable levels (for example API Standard 618).

4. CONCLUSIONS

Using PULTRAN it is possible to predict the minimum necessary pipe strength needed for safe operation of an installation and thus minimise the costs of piping. Furthermore, the effect and needed minimum respond time of safety-valves and relief valves can be checked in the design stage using this code. This way, damage to piping and rotating equipment due to high-amplitude waves can be prevented. The basics and possibilities of PULTRAN have been described in this paper by an example of a gas production system in which a pressure control valve breaks out.

REFERENCES

(1) Chorin, A.J., 1976, Random choice solution of hyperbolic systems, J. Comput. Physics, Vol. 22, p.517-533.

(2) Toro, E.F., 1987, The Random Choice Method on a non-staggered grid utilising an efficient Riemann solver, Cranfield College of Aeronautics, report no. 8708.

(3) Gottlieb, J.J., 1986, Lecture course notes on Random Choice Method for solving one-dimensional unsteady flows in ducts, shock tubes and blast wave simulators, AC Laboratorium Spiez, Switzerland, May 1986.

(4) Courant, R., Friedrichs, K.O., (1948), Supersonic Flows and Shock Waves, Interscience Publishers, Inc., New York.

(5) Collela, P., (1982), Glimm's Method for Gas Dynamics, SIAM J. Sci Stat. Comput. Vol. 3, No.1, March 1982.

(6) Looijmans, K.N.H., 1995, Homogeneous nucleation and droplet growth in the coexistence region of n-alkane/methane mixtures at high pressures, PhD-thesis, Technical University Eindhoven.

(7) Smolders, H.J., E.M.J. Niessen, M.E.H. van Dongen, (1992), The Random Choice Method applied to non-linear wave propagation in gas-vapour-droplets mixtures, Computers Fluids, Vol. 21, No. 1, pp. 63-75.

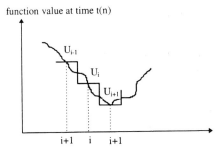

Figure 1 : Approximation of data at time t(n) by piecewise constant functions.

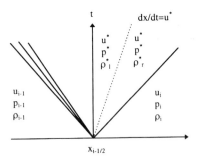

Figure 2 : Solution of the Riemann problem between grid points i-1 and i. In this example the left wave is a rarefaction and the right wave is a shock wave (Either wave can be a shock or expansion wave, allowing four possible combinations).

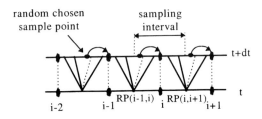

Figure 3 : Schematic illustration of Riemann solutions at the computational cells and the random sampling of the solution at a gridpoint.

C556/004© IMechE 1999

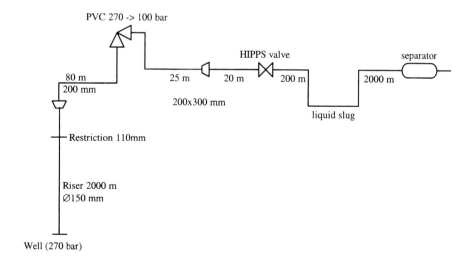

Figure 4 Simplified system lay-out

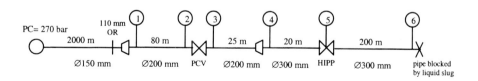

Figure 5 Lay-out of the simulation model

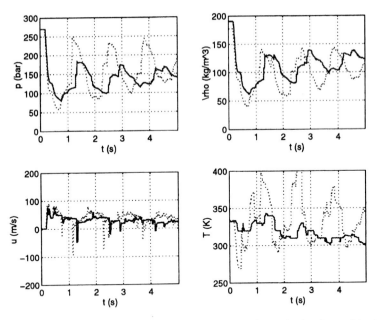

Figure 6 : Pressure, density, velocity and temperature as function of time after break out of the choke of the control valve. Position 1 (after the orifice) of figure 5. Dotted lines : ideal gas, no heat transfer, no wall friction. Solid lines : real gas, heat transfer and wall friction considered.

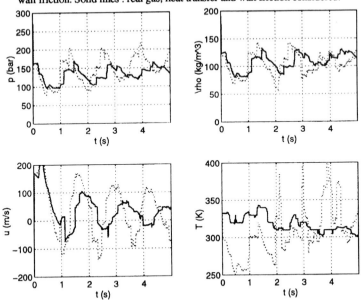

Figure 7 : Pressure, density, velocity and temperature after break-out of the choke in the pressure control valve. Position 2 of figure 5 (7 meters upstream of control valve). Line types according fig.6.

C556/004© IMechE 1999

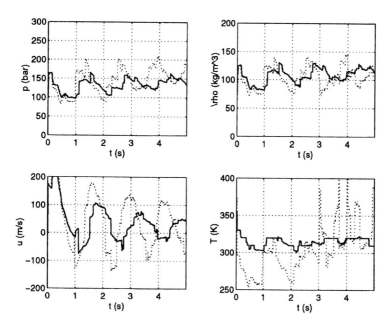

Figure 8 : Pressure, density, velocity and temperature after break-out of the choke in the pressure control valve. Position 3 of figure 5 (3 meters downstream of control valve). Line types according fig.6.

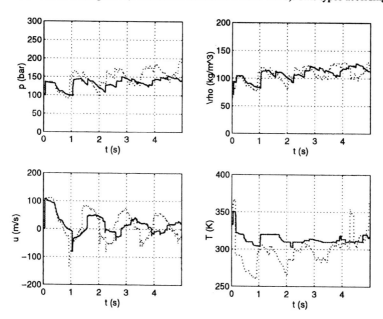

Figure 9 : Pressure, density, velocity and temperature after break-out of the choke in the pressure control valve. Position 4 of figure 5 (3 meters downstream the widthening). Line types according fig.6.

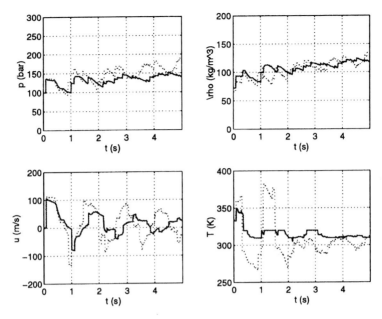

Figure 10 : Pressure, density, velocity and temperature after break-out of the choke in the pressure control valve. Position 5 of figure 5 (location of the HIPP valve). Line types according fig.6.

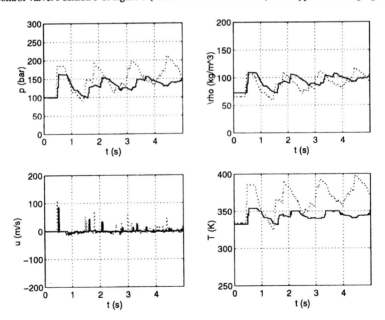

Figure 11 : Pressure, density, velocity and temperature after break-out of the choke in the pressure control valve. Position 6 of figure 5 (7 meters upstream of liquid slug). Line types according fig.6.

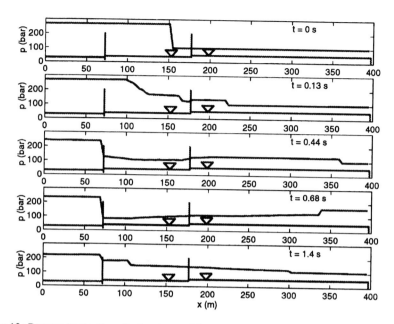

Figure 12 : Pressure in pipe line after break-out of the choke in the pressure control valve (The time of break-out corresponds to t=0). Line types according fig.6.

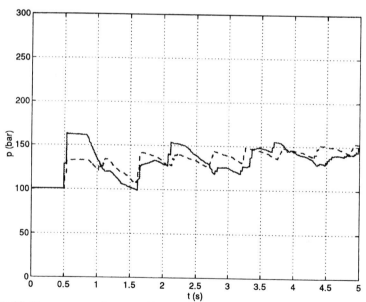

Figure 13 : Pressure versus time at position 6 of figure 5 for initial layout (solid line) and after installation of an orifice 109 m. downstream of the HIPP (dashed line). The HIPP is activated 2.1 seconds after the break out of the choke so the pressure does not exceed the allowable level of 150 bar.

C556/030/99

Benefits of high-speed integrally geared compressors in the oil and gas industries

D LAMY
Technical Support and Marketing, Sundstrand Compressors, Arvada, USA

ABSTRACT

Integrally geared centrifugal compressors continue to gain acceptance in process gas applications as an alternative to traditional overhung and between bearing centrifugal compressors, and even in some reciprocating compressor applications. When compared to traditional multistage centrifugal compressors, they can provide savings in initial capital cost by performing the same compression ratio with a smaller number of stages; furthermore, by mounting compressor, gearbox, drive train and all auxiliary systems on the same skid, footprint and field installation costs can be reduced.

When compared to reciprocating compressors, they offer higher reliability and savings in reduced downtime and maintenance costs. They also offer a great deal of flexibility in packaging configurations, to suit the specific requirements and challenges of each application.

Originally developed for air, Integrally Geared Process Gas Compressors have demonstrated the necessary high levels of safety, reliability and robustness, and today are routinely specified as an appropriate alternative to traditional API-617 compressors. Recognition of this fact is why a new chapter dedicated to Integrally Geared Centrifugal Compressors is being incorporated into the upcoming 7th edition of API-617.

CONTENTS:

A BRIEF HISTORY OF HIGH SPEED INTEGRALLY GEARED CENTRIFUGAL COMPRESSORS

While the benefits of high speed centrifugal compressors had already been recognized earlier in the century, the first Integrally-Geared high-speed centrifugal compressors appeared in the 1940's, and were exclusively designed for air duty.

In the late 1950's, a high-speed integrally geared centrifugal pump was developed by Sundstrand, whose purpose was to boost the power for the Boeing 707 jet engine during take off. This pump, which had to operate at relatively low flow rates and run dry for the duration of the flight, used a single, straight- radial-bladed impeller, and was an adaptation of the partial emission technology developed by U.M. Barske. Eventually, as airport runways became longer and the need for high take-off power subsided, the production of this pump was abandoned, but Sundstrand began a search for a niche in the industrial market for low flow, high pressure centrifugal pumps and compressors, that resulted in the commercialization, in the early 1960's, of the Sundyne High-Speed Integrally-Geared Pump in the hydrocarbon, chemical and petrochemical industries. The High Speed Integrally Geared Centrifugal Compressor, whose first appearance followed in 1965, was the first of its genre to be introduced in the process industry.

During the 1970's, the wide diffusion of the multi-stage horizontal barrel centrifugal compressor considerably expanded the area of application of centrifugal equipment to process gas duties. API Standard 617 *"Centrifugal Compressors for Petroleum, Chemical and Gas Service Industries"*, was written as a recommended practice for the design and manufacture of this type of centrifugal compressor, to ensure that the necessary features of safety, reliability and robustness required when handling process gas applications would be incorporated in it.

Today, high-speed integrally geared centrifugal compressors are part of the product range of every world-class manufacturer of centrifugal compressors. Over more than 35 years of successful field experience, Sundstrand alone has accumulated over 1600 installations of high-speed IG compressors in the most diverse process services, with molecular weights ranging from 2 (Hydrogen) to 70 (Chlorine).

C556/030/99

TYPES OF COMPRESSORS AND THEIR PERFORMANCE ENVELOPES

Compressor technology varies widely, to suit the many different performance requirements associated with the all but unlimited range of process duties. All compressor types, however, can be categorized in two major technological categories, i.e. that of "Positive Displacement" and "Dynamic" compressors. Positive Displacement Compressors are those in which isolated volumes of gas are subsequently displaced from a relatively low pressure region to a relatively high pressure region. The throughput of these compressors is the product of the quantity of gas contained in each of the isolated volumes times the number of volumes displaced per unit of time. The discharge pressure is determined by the pressure existing in the region where the isolated volumes of gas are displaced to, and remains essentially independent of the dynamic forces acting on the gas while it passes through the compressor.

Reciprocating (or piston) compressors are the most classic exemplification of this concept, but screw compressors, sliding-guide-vanes compressors, lobe compressors, gear compressors and liquid ring compressors are all based on the same principle.

In a "Dynamic" Compressor the gas stream is accelerated through the mechanical action of blades, thereby increasing its kinetic energy, which is subsequently transformed in static energy (increased pressure) through a controlled deceleration occurring in a "diffuser" section, aerodynamically designed to minimize energy losses. Centrifugal, axial and regenerative compressors all belong to the "dynamic" category.

To further increase the discharge pressure that can be obtained by a compression system, several compressors can be arranged in a "series" configuration of multiple compressor stages, where the outlet of each compressor stage is connected to the inlet of the subsequent stage. To increase the volume throughput handled by the system, the stages can be connected in "parallel", where the inlets of all the compressor stages are connected to a common suction manifold, and the outlets of the same stages are all connected to a common discharge manifold.

Positive displacement compressors are ideally suited for applications requiring relatively low volume flows and high pressure differentials, while the operating principle of dynamic compressors naturally lends itself to satisfy those situations where high volume flows and relatively low pressure differentials are requested.

The different type of action performed by positive displacement and dynamic compressors is shown in Fig. 1:

A) A positive displacement compressor will tend to deliver the same volume flow rate, irrespective of the pressure existing in the discharge region, provided the speed remains unchanged (the actual behavior is influenced by the "slip factor" - the re-expansion during the suction stroke of the gas that has remained trapped in the compression chamber at the end of the compression stroke - and backward leakage from the high-pressure region to the low-pressure region).

B) In a dynamic compressor, operating at constant speed, the "head" (energy transferred to the gas per unit of mass flow) generated by the compressor will tend to increase, decrease or remain constant with increasing volume flow rates, depending on the geometry of the blades. For example, in the case of a centrifugal compressor using straight radial blades, the head will remain constant irrespective of the volume flow it handles (actual operation will be affected by the friction losses, which tend to increase with the square of the speed of the gas stream through the compressor, and by the incidence losses, related to the angle of attack of the gas at the impeller inlet).

Fig. 2 schematically illustrates the typical performance envelopes of the main compressor families. The figure shows how the performance envelope of Integrally Geared Centrifugal Compressors overlaps that of many other technologies, often representing a viable alternative to maintenance-intensive reciprocating compressors on the low-flow end of its range, as well as to initial-capital-intensive multistage centrifugal and axial compressors on the opposite, high-flow end. Through the incorporation of Barske's "partial-emission" design, Sundstrand's compressors have pushed the limits of the "high-speed integrally-geared" technology towards the low-flow end of their envelope, invading the traditional area of single-stage reciprocating and rotary compressors.

Fig. 3 shows the typical curve of a centrifugal compressor, and its operating range. At the lower flow end, the curve's useable range is limited by the "surge" phenomenon, represented by a stalling of the gas, or flow separation from the blade profile, where the gas is partially recirculated back to the impeller's suction, giving rise to oscillating (backwards and forwards) flow cycles accompanied by suction temperature increase, high noise levels and potentially destructive mechanical vibration. At the opposite end of the curve, at high flow rates, the "choke" limit is the flow rate region where the incidence angle of the gas on the inlet vanes becomes high enough to reduce the entrance flow area and force the Mach number (ratio between the gas velocity and the acoustic velocity through the gas stream) up to the point where the volume flowrate through the compressor cannot be further increased.

BENEFITS OF HIGH SPEED INTEGRALLY GEARED CENTRIFUGAL COMPRESSORS

The primary advantage of high speed is that, by increasing the kinetic energy imparted to the gas stream by the impeller vanes, it allows to develop the same amount of head using a lower number of compressor stages. This, in turn, reduces both the initial capital investment, through a reduction in footprint size, and the operating costs, through a reduction in spares inventory and maintenance costs.

One way to obtain a higher rotating speed is through the inclusion of a gearbox between the driver and a multistage compressor. With this arrangement, the various stages of the compressor are all mounted on the same shaft, which now rotates at a speed equal to the gearbox output speed.

The fundamental advantage of Integrally-Geared centrifugal compressors (Fig. 4) with respect to the separate-gearbox-type configuration is that the various stages can be mounted on separate pinions, all driven by a common low-speed bull gear; the pinions can rotate at different speeds, thereby allowing to optimize each stage's rotating speed and dramatically increasing the flexibility of the design.

The specific advantages derived from this increased flexibility are multiple:

A) First and most important, the efficiency of each stage can be maximized through the appropriate selection of its operating speed, thereby lowering the power consumption of the compressor;
B) The greater flexibility in the choice of each stage's operating speed enables to use speeds up to 5 times higher than those typically used with conventional barrel type centrifugal compressors using separate gearboxes. It is not uncommon today to find integrally geared centrifugal compressor applications running at 60,000 rpm and higher. With this increased output speed capability, it is possible to maximize the head obtainable from each stage. For example, integrally geared technology enables to obtain pressure ratios of 10:1 in two stages where a conventional horizontal, split case design would have to use 8 stages;
C) The fact that the various impellers are no longer mounted next to each other on the same shaft, but are rather mounted at the extremity of separate pinions, each with its own casing and suction/discharge connections allows to optimize not only the speed of each impeller but also its design and its dimensions, as well as the inlet and outlet flow conditions to each stage. For example, it is much easier, with an integrally geared compressor, to equip each stag with a set of adjustable inlet and outlet guide vanes, for those services where the convenience to efficiently meet off-design conditions warrants for the increased complexity of the system. A typical situation of this kind is that of air compressors operating in services characterized by widely varying ambient air temperatures;
D) For the same reason, it is easier to install coolers at the outlet from each stage, to decrease the discharge temperature of the gas stream exiting that stage, thereby increasing the thermodynamic efficiency of the subsequent stages;
E) The increased speed of each impeller generally translates to smaller dimensions both for the impeller and for its casing, and this results in a more economical use of exotic materials for those services which should require it;
F) The compressor can be packaged in a more compact and integrated design, offering further savings in space consumption and capital investment; furthermore, a single vendor supplies the compressor, the gearbox, and the relevant auxiliary systems, greatly simplifying the order execution process and communication interfaces;
G) The scope of supply is simplified through the elimination of high speed couplings. This, in turn, improves the reliability of the compressor package, reducing maintenance costs as well as spare parts inventory.

All of the above mentioned advantages concur to lower both initial capital investment and operating costs, resulting in a total life cycle cost saving.

TECHNICAL CHARACTERISTICS

Having seen the general advantages related to the use of integrally geared high speed centrifugal compressor technology, we will now focus a little closer on the detail of some of the technical characteristics that make the use of that technology possible.

A. High Performance Impeller Design

Two fundamental elements contribute to today's dramatic technological improvements in the performance of high speed impellers, and these are the use of sophisticated computational fluid dynamic modeling software on one hand and of 5-axis, numerical control milling technology on the other.

Computational Fluid Dynamic (CFD) modeling allows to optimize the aerodynamic design of high-speed impellers for each specific application, enabling to design 3-D blade profiles (Fig. 5). The software available is quick and user-friendly, and virtually removes all limitations to the optimization of each impeller's design. On the other hand, the advantages of using 5-axis milling with NC machines as a manufacturing process enables to translate the increased design capability into actual products (Fig. 6).

Prior to the advent of this technology impeller designs were limited by casting models available. Each new design required the manufacture of a new casting model, with a high impact on both cost and time. The new 5-axis, NC milling process enables to obtain higher fidelity of the actual product to the design, greater strength, less porosity, the elimination of casting defects, better surface finish and repeatability.

A further advantage is that many exotic materials are easier to machine than to cast, and this widens the range of materials that can be used for difficult applications, to meet the specific nature of the gas being processed, or that application's specific rotordynamics and critical speed considerations.

The benefits of the above features are improved efficiency gain, reduced power consumption, lower temperature rise and greater performance coverage, with higher turndown ratios.

B. High-Pressure Capability

The challenge posed by many process gas applications is that of high pressures, often accompanied by the hazardous nature of the process gas, which demands a stringent control of leaks to the environment and within the system.

Two are the fundamental features that allow today's integrally geared centrifugal compressors to meet this challenge, namely their rugged design and the development of low-leaking dry (gas) mechanical seals, now available in many different types and configurations. Before the appearance of dry gas mechanical seals, the use of oil seals obligated compressor manufacturers to equip their products with expensive auxiliary systems for the separation of the seal oil from the process gas which would inevitably contaminate it.

Over the years, since the first introduction of High Speed Integrally Geared Centrifugal Compressors in the Process Gas Industry, the dry (gas) mechanical seal technology has evolved to provide seal designs capable of handling up to 275 BAR (4,000 psi) pressure differentials, in single, tandem (unpressurized) and double (pressurized, with common rotating ring) configurations, as well as to equip them with a wide variety of seal environment control systems to suit the most diverse process situations.

Although the number of possible mechanical seal designs and configurations is as high as that of possible compressor types, all mechanical dry (gas) seals are based on the concept of two opposed faces, one of them being stationary (attached to the compressor's casing), and the other rotating with the high speed shaft (Fig. 7). The two faces are designed with different hardness degrees, so that it will always be the same face that wears out and needs periodical replacement. During the compressor's operation, the two faces are separated by a thin film of gas, which prevents them directly contacting each other, thereby reducing friction, wear and mechanical losses. The two faces are pushed towards each other by the action of springs, increasing the pressure in the gas film between them and reducing the thickness of the same film, as well as its leakage rate through the seal.

The actual design of the seal varies widely with the seal manufacturer and the specific application requirements. In a common type, shown in the left half of Fig. 8, the stationary face, in carbon graphite, is designed with radial grooves. The surface of the face between two adjacent grooves is in relief, creating the hydrodynamic pad against which the gas film operates. Any solid particulates that might be entrained in the seal gas will be evacuated through the radial grooves, preventing them to damage the surface of the seal. This type of design has demonstrated to be especially tolerant of "dirty" seal gas (seal gas entraining solid particulates), at the expense of a relatively high seal gas leakage rate through the seal.

In another seal design (right half of Fig. 8) it is the rotating face, commonly in tungsten carbide, that is carved with spiral grooves. The seal gas flows from the outer region of the seal along the grooves, and as these narrow toward the inner diameter, the gas film pressure increases, forcing the two faces to separate from each other, against the force of the springs. This type of design has proven to reduce considerably the seal gas leakage rate with respect to the radial groove design (leakage rates of approximately 1 Kg/h - 0.037 lb/min - are common with this design), but it does so at the expense of a lot lower tolerance to the presence of particulates in the seal gas stream. It becomes necessary with the spiral groove design seal (unless the seal gas stream is especially clean) to include 5-μm filters upstream of the inlet to the seal region.

The latest development introduced in the field of dry seal technology is the "wavy face" rotating seal design (Fig. 9), where a stationary carbon ring faces a rotating Silicon Carbide ring designed with a wave-shaped profile all around its circumference. The waviness is most pronounced at the outer circumference of the rotating face, and gradually reduces along its radius, until it totally disappears at the inner circumference of the seal. The effect of the wavy-shape of the rotating face (Fig. 10) is that, after entering the rotating face's surface at the lower portion of the wave, the seal gas is pushed by the rotating movement of the face towards the high part of the wave profile, thereby increasing its pressure and forcing the separation of the stationary face from the rotating profile. The combined effect of the shape of the wave and the increased pressure forces the bulk of the seal gas stream at this point to redirect itself outward, in this way greatly reducing the amount of seal leakage through the seal, without decreasing its tolerance to the presence of particulates, as these are entrained by the gas recirculated to the outside of the seal.

This design has been proven to work efficiently in applications up to 34,000 rpm and 85 BAR (1,230 psi) pressure differentials. It is suitable for both liquid, gas and mixed seal fluids. It is available in cartridge configurations, to meet the requirements of API Standard 682.

Among the benefits of the wavy face seal technology, the most important are:
A. Bi-directionality of the wave design, which allows the seal to run backwards with no problem;
B. Smoothness of the wave profile, which reduces potential wear and damage in case of contact (e.g. during the compressor's start-up);
C. Lower leakage rates through the seal;
D. High pressure and high speed capability (up to 85 BAR - 1,230 psi -, 34000 rpm);
E. Resistance to contamination from solid particles entrained in the seal gas stream or from oil particles dripping from the integral gearbox in vertical compressor configurations, in case of oil seal failure or excessive wear;
F. Possibility to retrofit existing installations with little or no modifications to existing hardware.

C. Partial Emission Technology
On the lower end of the specific speed range, the adaptation of Barske's partial emission design to centrifugal compressor technology has enabled Sundstrand to further increase the flexibility of integrally geared centrifugal compressors, by designing them to operate at lower flows than any conventional dynamic compressor had ever been able to handle before.

The result is that for many process applications characterized by low-flow and high-head, that used to be the traditional realm of positive displacement compressors, customers have today the possibility to benefit from the high-reliability, pulsation free performance of a dynamic compressor. There are multiple advantages associated with the use of a centrifugal compressor as opposed to a positive displacement compressor, the most important of which can be summarized in:

- pulsation-free operation (eliminating the need for pulsation dampening equipment)
- oil-free gas stream (no contamination of lubricating oil in the gas stream, and elimination of the need for oil/gas separation equipment downstream of the compressor's outlet):
- higher reliability, and lower maintenance costs;

- elimination of the need for pressure safety devices to protect the system from excessive pressures.

These advantages are often more than enough, in the eyes of end-users, to compensate the typically lower efficiency of a centrifugal compressor, when compared to a reciprocating compressor.

Partial emission technology (Fig. 11) is based on the use of a single point emission from the impeller outlet: the fluid stream is conveyed, through a diffuser throat, to a diverging duct where its kinetic energy is converted to pressure, as opposed to the full emission volute design, where the diffusion process takes place all around the impeller. The geometrical configuration of the compressor casing around the impeller is concentric, and the impeller itself is a simple open, straight-radial-bladed design. For centrifugal compressors, this technology has proven its superiority in the low flow (100 to 800 m^3/h - 60 to 500 ft^3/min) and high head (up to 85,000 N-m / Kg - 28,000 ft-lb$_f$/lb$_m$) region.

When coupled with the high speed integral gear concept, partial emission technology will use ample clearances between the impeller and surrounding surfaces, eliminating wear and originating a performance that will never decay during the compressor's operating life. Fig. 12 shows that for very low specific speeds (low flowrates, high heads) the use of the partial emission concept results in higher efficiencies than those obtainable with state-of-the-art full emission designs.

INDUSTRY STANDARDS

When dealing with process gas applications, the issue of compliance to industry standards is often a crucial one, that can easily throw a competitor out of the game, if not examined in due detail and with the necessary background of knowledge and critical thinking. The best approach is always to penetrate the mere face value of each requirement, to understand the ultimate concerns that lie behind it and that have prompted it. It is an answer to those essential concerns, and not just the compliance to the literal wording of the requirement, that is needed to safely and reliably handle challenging and hazardous process applications.

This is particularly true with High Speed Integrally Geared Centrifugal Compressor technology. In fact, such world-wide utilized industry standards as API-617, *Centrifugal Compressors for Petroleum, Chemical and Gas Service Industries*, API-613, *Special Purpose Gear Units for Petroleum, Chemical and Gas Industry Services* and API-614, *Lubrication, Shaft-Sealing and Control-Oil Systems for Special Purpose Applications*, were all conceived before the appearance on the market of high-speed integrally geared technology, and were not addressed to this technology. This consideration is at the root of the virtual impossibility for an integrally geared high speed centrifugal compressor to *literally* comply with each and all of the requirements of these standards. Still, integrally-geared high speed centrifugal compressors operating in process gas applications can and should be designed and manufactured to meet the fundamental intent of the above standards, which is primarily focussed on reliability and safety issues, and secondarily with testing, instrumentation and vibration issues.

API-617, for an example, was originally conceived to address conventional between-bearing, horizontal barrel multistage centrifugal compressors running at low speed, or using separate speed-increasing gearboxes which could step their speed up to levels seldom exceeding 12,000 rpm. Today's integrally-geared centrifugal compressors can be designed for rotating speeds up to five times that level, and have a radically different layout concept, posing challenges and offering benefits that are understandably of a radically different nature: high speed couplings are not an issue with IG compressors; torques are generally much lower; space, on the other hand, is at a premium, and the same degree of redundancy in auxiliary apparatus that did not constitute an issue with conventional centrifugal compressors, may be hardly adaptable to compact, space saving packages.

Still, integrally geared centrifugal process gas compressors have by now been on the market for more than 35 years, proving that they can provide a reliable solution for hazardous and difficult process services. In other words, they have proven that, when properly designed and packaged, they do meet the fundamental requirements of reliability, safety and robustness which API-617 considers necessary features to handle process gas applications. In fact, since their first appearance, high-speed integrally-geared centrifugal compressors have conquered an ever-increasing share of the process gas application market, eroding that of competitive technologies such as reciprocating compressors and multistage barrel centrifugal compressors, and thus have gained industry recognition. This recognition also comes today from the American Petroleum Institute itself, and this is why a sub-committee of API is presently working at the next (7th) edition of Standard 617, which will incorporate a chapter specifically dedicated to IG high speed centrifugal compressor designs.

Concerning API Standard 613, it must be observed that, due to their history, Sundstrand Compressors' gears were developed as an extension of Sundstrand's aerospace know-how, and followed aerospace standards rather than industrial (API, AGMA) standards. Their gear ratings compare reasonably well with AGMA gear ratings, meeting their objectives of life and reliability. API-613, on the other hand, was written for gears designed to rotate at much lower speeds than those attained by today's high-speed integrally geared designs, where torques and tooth bending were the primary concerns, and doesn't seem to recognize the necessary gear differences at high speeds or gear pitch line velocities. Its approach consequently leads to extremely conservative design requirements that can actually become harmful at high pitchline velocities. This is where integrally geared, high speed centrifugal compressor manufacturers can experience a design conflict, when trying to force the design of their product into compliance with API-613 requirements.

As for API Standard 614, this standard was also written for conventional centrifugal equipment, using separate auxiliary systems, for which space consumption was not a primary concern. Its basic concept is redundancy in the use of auxiliary protections, instrumentation and components, designed to reach a higher level of reliability. But this idea conflicts with one of the basic advantages of high-speed integrally geared designs, i.e. the compactness and integrated packaging that they afford to produce. Particularly in the case of very low power centrifugal compressors, it is common to find that an auxiliary lubrication system designed in complete accordance with API-614 will cost more than the compressor itself. On the other hand, 35 years of field experience with simpler lube systems, designed based on the manufacturer's recommendations, have often proved to be just as reliable as an API-614 system. In these cases (very low power compressors, up to 400 KW), it might be a wiser approach to require the integrally-geared centrifugal compressor manufacturer to produce a list of good references than to force him to full compliance with a standard that was not originally written for the type of equipment he manufactures.

MARKETS AND APPLICATIONS

The use of IG centrifugal compressors is widespread in the process gas industry, and it would be impossible to list all of the services and applications where this technology is conveniently applied.

Still, a quick overview of existing process gas installations shows that integrally geared centrifugal compressors have been employed in the Refinery, Petrochemical, Chemical, Power Generation and Natural Gas Production and Extraction Industries.

In the Refinery industry, the characteristic combination of low-flow and high-suction-pressure capabilities has made high-speed integrally geared partial emission centrifugal compressors a typical choice for such services as Hydrogen feed and Hydrogen Recycle, used to prevent the deposition of coke on the catalyst and the formation of olefins; Lift Gas Blower is another application, commonly featured in the regeneration section of platforming units to regenerate the spent catalyst with a Nitrogen stream, for which licensors routinely specify integrally geared centrifugal compressors. Other typical refinery applications for this technology are Vapor Recovery, Tail Gas and H_2S boost, all commonly present in Hydrocracking, Hydrotreating, Platforming and Isomerization units.

In the Petrochemical industry, common IG compressor applications are Ethylene and Propylene Recycle, where unreacted Ethylene and Propylene vapors are recompressed, passed through a heat-exchanger to bring them back to reaction temperature, and returned to the reactor to be re-processed, to maximize the plant's output. Hydrogen Recycle is a common application also in these plants.

In the Natural Gas Production and Extraction Industries, both On- and Off-Shore, integrally geared compressors can be conveniently employed for such services as Natural Gas Regeneration (the gas is continuously recompressed, reheated and passed through the dryer being regenerated - while a second dryer remains in operation - to pick up the liquids adsorbed on the catalyst's surface, carrying them to a condenser, where they are separated from the gas stream). Refrigeration, where cold propane is compressed to act as a refrigerant for the main gas stream, to recover the heavier hydrocarbons fractions (propane, ethane and butanes) in the liquid form, is another common IG-centrifugal compressor application in the Gas Production Industry. Gas Gathering, where a single compressor must have the flexibility to handle gas arriving from several separation units with different flowrates and suction pressures is yet another such application. Other common applications are Gas Lift (the gas is reinjected in the well to lower the specific gravity and the viscosity of the oil and thus facilitate its recovery from the reservoir); Gas Re-Injection (to maintain the reservoir's pressure, compensating for its depletion, thus prolonging its productive life); Main Transport Boost (from the production plant to the nearest Gathering or Processing Center); and such Gas Turbine support services as Fuel Gas Feed, Air Atomization (to reduce the NO_x levels generated by the combustion) and Turbine Blade Cooling.

Typical services performed by IG compressors in Power Generation plants are Fuel Gas Boost, Air Atomization and Gas Turbine Blade Cooling.

In the Chemical industry, Regeneration and Recycle, together with Ammonia, Chlorine feed, Acetylene, Styrene and HCl feed are but a few of the applications where the use of integrally geared centrifugal compressors is commonly adopted.

CONCLUSIONS

Integrally geared turbo machinery can be applied on a wide range of process gas applications, offering lower capital investment and lower operating costs, together with high reliability, and great packaging and configuration flexibility.

Use of this technology in critical process gas duties is rapidly growing and is prevented from growing even faster only by the uneasiness with which a relatively new technology is always perceived in an established industry. Even so, the technology has now won a high level of recognition by every major actor in the industry worldwide, from end-users to engineering contractors, manufacturers and international standard institutions.

Within this growing market niche, over 35 years of field experience, Sundstrand has developed and affirmed its capability to apply IG high speed centrifugal compressor technology to low-flow, high-head, high-suction-pressure applications, where reciprocating or other positive displacement compressors used to be the only available option.

At time of publication no figures were available for this paper.

Authors' Index

SUNDSTRAND
COMPRESSORS
THE Low Flow Centrifugal

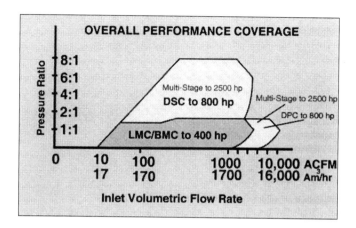

Sundstrand Compressors (SC) has over 30 years experience in process gas applications. SC specializes in low-flow, integrally-geared, high-speed centrifugal technology. SC broad range of products overlap many positive displacement designs including reciprocating and rotary screw type compressor packages. Benefits and advantages of the SC integrally-geared packages include the following:

- High performance custom impeller designs
- Reduced overall weight and space requirements
- Lower installation and foundation costs
- Smooth pulsation free flow
- Non-deteriorating performance
- Longer mean time between maintenance
- Oil free gas compression
- Mechanical dry gas sealing

SC has a worldwide sales, manufacturing and support network, providing the highest level of customer satisfaction. Next time you are in the market for a process gas compressor, in the ranges pictured above, check out SC. SC delivers excellent value, delivery and service at a competitive price!

North /South American Sales:
Golden, Colorado
Phone: 1-303-425-0800
Fax: 1-303-940-2808

International Sales:
Dijon, France
Phone: +33-3-80-383300
Fax: +33-3-80-383371

Or visit Sundstrand Compressors on the world wide web at www.sfh.com

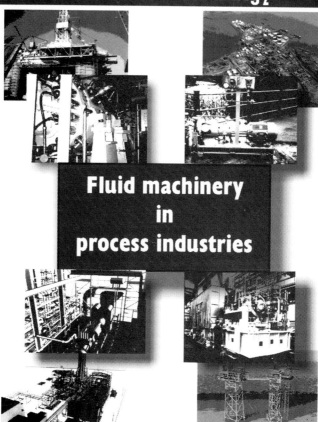

SULZER Roteq Group

Sulzer Pumps

Sulzer Turbo

Sulzer Burckhardt

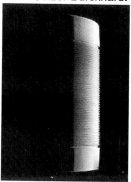

Innovation through Excellence in Technology

Recent Achievements (1990's)

- Development of multiphase pump
- Development of oil-free compressor (HOFIM)
- Introduction of pipeline compressor (MOPICO)
- First horizontal labyrinth-piston compressor
- First labyrinth-compressor with trunk piston
- First vertical injection pump installed

SULZER Roteq Divisions:

Sulzer Pumps Ltd
CH-8404 Winterthur
Tel. +41/52-262 11 55
Fax +41/52-262 00 40

Sulzer Turbo Ltd
CH-8023 Zürich
Tel. +41/1-278 22 11
Fax +41/1-278 29 89

Sulzer Burckhardt Ltd
CH-8404 Winterthur
Tel. +41/52-262 55 00
Fax +41/52-262 00 51

.